Indoor Environment and Sustainable Building

Series Editors

Angui Li, Xi'an University of Architecture and Technology, Xi'an, China

Risto Kosonen, Department of Mechanical Engineering, Aalto University, Vantaa, Finland

The book series "Indoor Environment and Sustainable Buildings" publishes insights and latest research results on indoor environment and sustainable building, and aims to provide a more energy-efficient, safer and healthier space engineered for human to live and work. The intent is to cover all the technical contents, applications, and multidisciplinary aspects of the engineering techniques for improving indoor environment quality and the energy efficiency of heating, ventilation, and air conditioning (HVAC) systems, as well as the smart home devices, sustainable architectures and buildings.

Topics in the book series include:

- indoor environment
- thermal comfort
- heating, ventilation, and air conditioning (HVAC) systems
- smart home devices
- energy efficiency
- building physics
- building services
- sustainable architectures and buildings, etc.

The objective of the book series is to publish monographs, reference works, selected contributions from specialized conferences, and textbooks with high quality in the field of indoor environment and sustainable building. The series provides valuable references to a wide audience in the community of HVAC researchers, indoor space designers, policy makers and architects.

Tianyi Zhao · Jiaming Wang · Yiting Wang

Thermal Distribution Principle and Nonlinear Optimal Control for HVAC Water System

Springer

Tianyi Zhao
Department of Civil Engineering, School of Infrastructure Engineering
Dalian University of Technology
Dalian, China

Jiaming Wang
Department of Civil Engineering, School of Infrastructure Engineering
Dalian University of Technology
Dalian, China

Yiting Wang
Department of Civil Engineering, School of Infrastructure Engineering
Dalian University of Technology
Dalian, China

ISSN 2730-7042 ISSN 2730-7050 (electronic)
Indoor Environment and Sustainable Building
ISBN 978-981-96-0376-3 ISBN 978-981-96-0377-0 (eBook)
https://doi.org/10.1007/978-981-96-0377-0

This Springer imprint is published by the registered company Springer Nature Singapore Pte Ltd.
The registered company address is: 152 Beach Road, #21-01/04 Gateway East, Singapore 189721, Singapore

Foreword

On the day of the Paris Olympic opening, I received Prof. Tianyi Zhao's significant work on thermal distribution and optimal control for HVAC water systems. After went through this interesting book, I am delighted to introduce it to researchers, professionals, and students!

This comprehensive book begins with modeling the primary components of HVAC water systems, delving deeply into the thermal distribution characteristics of systems and elucidating the identification methods for the most unfavorable loops. Methods for enhancing the identification of hydraulic and thermal distribution characteristics of pipeline networks are introduced for differential pressure resetting, thereby significantly improving system operation and control robustness. In addition, the book addresses the lag response characteristics of system parameters and the nonlinear operating characteristics of components, providing detailed modeling methods to identify and quantify the key factors influencing these lag characteristics. Based on these modeling efforts and characteristic analyses, the book presents several useful global optimization methods for optimal scheduling and setpoint determination. A comprehensive evaluation of these optimization methods, in terms of system decoupling effects, control efficacy, and energy efficiency, is also provided.

One of the most impressive features of this book is its systematic introduction to the operational characteristics of HVAC water systems, control loops, parameter distribution characteristics, optimization methods and algorithms. The content is presented in a manner that is both comprehensive and highly readable. The methods and algorithms introduced represent the cutting-edge technologies, offering practical guidance for optimizing real-world systems.

This book is an invaluable resource for researchers, educators, and students in the field of heating, ventilation, and air conditioning. The principles and engineering solutions provided will aid in advancing academic and practical knowledge for researchers and professionals. As a culmination of Prof. Tianyi Zhao's extensive efforts, this book makes a significant contribution to the field, offering both theoretical insights and practical solutions. I highly recommend this book to anyone involved with HVAC systems, including researchers, educators, design professionals, and facility management experts. It will provide an in-depth understanding and updated performance enhancement methods for the important air-conditioning system, i.e., water systems.

August 2024

Prof. Shengwei Wang
Chair Professor of Building Energy and Automation
Otto Poon Charitable Foundation Professor in Smart Buildings
Director of Research Institute for Smart Energy
The Hong Kong Polytechnic University
Kowloon, Hong Kong

Preface

Since around 2005, the Chinese government and academic community have started to focus on the energy consumption monitoring and energy-saving diagnosis of large public buildings. From public institution buildings to commercial office buildings, and then to campus buildings and hospital buildings, the continuous policy support, standard and norm formulation, and demonstration project construction have achieved a series of scientific research and engineering practice results. These achievements provide valuable references for the energy-saving diagnosis, renovation, and performance analysis of public buildings in China and worldwide.

I have been fortunate to experience these development stages since my master's studies and have deeply participated in related scientific research projects and engineering practices. I have led or participated in on-site testing, energy auditing, and energy-saving renovation of nearly 500 public buildings' energy systems. Throughout this process, I deeply realized the difficulties and obstacles in achieving online, global control of building energy systems, especially HVAC systems. Therefore, in 2006, I closely combined the needs of on-site testing and engineering practice and chose the energy-saving optimization control of air-conditioning water systems as the research direction of my doctoral thesis.

It has been nearly 20 years since I first came into contact with the optimization control of HVAC systems. This direction has become the most important mainline of my scientific research and teaching work development, and I have cultivated more than 10 master's and doctoral students in the relevant field. I feel it is necessary to organize the phasic research findings into a monograph to share with students, researchers, and engineers dedicated to related fields.

Taking the air-conditioning water system with particularly complex nonlinear characteristics as an example, this book aims to explore three questions from the perspective of engineering practice with the readers: (1) What is the reference object for dynamic control of the air-conditioning water system, and what are the differences between the most unfavorable loop in the design stage and the operation stage? (2) How to build control loops for online applications, comprehensively considering system efficiency and regulation stability to construct practical and efficient control strategies and algorithms? (3) How to understand the transmission characteristics of quality parameters and quantity parameters in the pipe network from the perspective of nonlinear characteristic identification and then propose universal control methods to address the challenge of global optimization control of air-conditioning water systems.

The content of this book can be divided into three parts: (1) Chapters 1–3 introduce the research motivation and progress, defining the research object and its model; (2) Chapters 4 and 5 define the concept of the most unfavorable thermodynamic loop, clarify the reference object of online optimization control, and propose a series of differential pressure set value and pump optimization control methods from the perspectives of pipe network stability and online control needs; (3) Chapters 6 and 7 theoretically analyze the nonlinear variation characteristics of control parameters in the pipe network of the air-conditioning water system and, based on this, propose globally optimal control methods with excellent operability and universality.

Upon the completion of this book, I would first like to express heartfelt gratitude to Prof. Jili Zhang from Dalian University of Technology. He has been not only my mentor from undergraduate to postdoctoral studies but also a guide throughout my academic and teaching career. From being teacher and student to colleagues, we have worked together for over 20 years. The cultivation and assistance provided by Professor Zhang have been the most important support for this book. Secondly, I would like to thank my students, especially the co-authors of this monograph, Dr. Jiaming Wang and Dr. Yiting Wang, who have devoted a great deal of effort to the manuscript of this book. Lastly, I want to thank Prof. Shengwei Wang from Hong Kong Polytechnic University. As a top international expert in the field of intelligent control of building energy systems, I am honored to invite Prof. Wang to write the foreword for this book.

Due to the author's limited capability, I earnestly ask readers to provide corrections or engage in discussions on the content or derivative directions of this book.

Dalian, China
July 2024

Tianyi Zhao

Contents

Chapter 1
System Overview

1.1 Research Background

The heating, ventilation, and air-conditioning (HVAC) system undertakes an extremely important task in the regulation process of indoor temperature and humidity environment of large public building, which accounts for more than 50% of the total energy consumption of buildings and more than 20% of the total society energy consumption in China. The HVAC water system, as an indispensable part of the HVAC system, accounts for up to 90% of system energy consumption, which has great energy conservation space. Therefore, the development of intelligent energy conservation techniques of HVAC water system for improving the overall operation energy efficiency is of great significance to global energy conservation and realization of the "carbon peaking and carbon neutrality" goals.

In recent years, with the rapid development of science and technology, the automatic control technique has become the most effective optimization method for energy conservation of HVAC water systems. However, due to the strong temporal variability, the complex spatial structure, and the coupling characteristics among control loops, the optimal control effect of practical system is not ideal. Some issues like unreasonable control parameter selection, inappropriate control loops' setpoints, and improper optimization timing are still common in the existing HVAC water systems in public buildings, resulting in hard decoupling control among control loops, which seriously affects the actual control effect and operating energy efficiency of the HVAC water system.

In order to investigate the energy consumption status, operation problems, and energy saving potential of HVAC systems in large public buildings thoroughly, many researchers have carried out the energy status investigation and field testing of large public buildings. Since 2005, the authors' research group has investigated and tested more than 150 large public buildings (including 7 public buildings in Harbin, 49 government buildings in Shenyang, 77 public buildings and government buildings in Dalian, 7 government buildings in Beijing, 2 public buildings in Suzhou, 2

T. Zhao et al., *Thermal Distribution Principle and Nonlinear Optimal Control for HVAC Water System*, Indoor Environment and Sustainable Building,
https://doi.org/10.1007/978-981-96-0377-0_1

public buildings in Shenzhen, and 8 public buildings in Omaha and one in Lincoln). According to field investigation and testing, it was found that the following serious operating issues exist in HVAC systems of large public buildings:

(1) The temperature setpoint of conditioned room is commonly set as 23–25 °C in summer, and the actual room temperature is generally lower. In winter, the opposite is true. Unreasonable temperature setpoint of conditioned room not only increases system energy consumption (1 °C increase of room temperature in summer can save more than 20% of energy), but also does not conform to the objective law of human thermal comfort.
(2) The actual efficiency of supply fan in HVAC air system is generally about 40%, which is relatively low. The supply air volume under partial load is too large, which sometimes may exceed more than 1 times. In the transition season, the regulation of fresh air volume is shortage, and the outdoor free cooling is rarely used, which increases the energy consumption of artificial cooling. The supply air ducts lack of hydraulic balance, which renders uneven heat distribution among conditioned rooms.
(3) The operating issues in practical HVAC water system are serious. The temperature difference between supply and return water of chilled water is generally about 3 °C, or even lower. The temperature difference between supply and return water of hot water is generally more than 40% lower than the designed temperature difference, even only 1–3 °C, which renders the huge power consumption of hot water circulating pump. The pipeline hydraulic imbalance issue is serious, and the pump efficiency is often less than 50%, which is a common phenomenon in some practical systems. For the parallel chillers, the problem of bypass operating issue of shutdown chiller is common.
(4) The automatic control issues of HVAC system are extremely serious. In the HVAC systems with automatic control techniques, almost all of the systems do not operate in accordance with the automatic control mode. The phenomenon of no monitoring, no control, and no use of automatic control system is very common in most of practical system. A few can realize only measurement but no control, and these systems cannot realize the important potential of automatic control techniques for optimizing and regulating the system. The control system lacks of commissioning process after implementation. In the recent years, after the energy conservation transformation of the pumps in practical systems, there are also some problems can be found in the transitional season, such as less energy conservation for pumps and insufficient pump head.

The large-scale public HVAC systems in the US are representative in the world, which has few mechanical issues. After the completion of the system, there is a balance debugging process for air volume and water volume (completed by professional Balance company), and the automation degree of the HVAC system is almost 100%. However, there are also some serious operating issues of HVAC systems in US:

(1) In the aspect of HVAC air system, the damper position of variable air volume (VAV) terminal box is not set correctly, and the problem of unexpected control

of VAV box is common. The static pressure of the primary air side of the VAV box is too high, resulting in frequent start and stop of the terminal fan, high power consumption, and even damage the fan motor. In the dual duct system, the problem of heat and cold offset is also very common.

(2) In the aspect of HVAC water system, the chilled water bypass on the side of chiller and cooling coil are relatively common issues. In addition, the control effect of regulating pump frequency is not ideal, and there are many issues during the operating process of water system, for example, when the regulating valve opening of the farthest user is small, there is an issue that the regulating valve opening of the middle branch user is fully opened but the indoor temperature still continues to rise.

(3) In terms of control system, the first common issue is the sensor failure, especially the humidity sensors and the differential pressure sensors, which would directly lead to the poor control effect of enthalpy-based air handling unit (AHU) and differential pressure based fresh air exhaust system. The second issue is that regulating valve actuator lacks of control ability correction after the installation, resulting in the loose actuator shaft and some other problems, leading to the correct control signal, cannot get the correct adjustment results. Thirdly, the improper control technique is also an important reason of poor control effect. The fixed parameter PI control is still a commonly used control algorithm, and all kinds of parameter setpoints (such as supply air static pressure, supply air temperature, chilled water differential pressure, and chilled water temperature) are unreasonable and lack of necessary resetting process. The selection of reference control variables for each controlled object, the coupling control strategies among multiple controlled objects, and the control logic of seasonal switching condition are unreasonable. Along with the VAV air-conditioning technique and pump variable frequency energy conservation technique, the above HVAC issues in US are also existed in China.

The above issues lead to high energy consumption and serious waste of resources in existing large public buildings. Some HVAC system with high energy consumption may guarantee the terminal and indoor air-conditioning demand, but there are quite a few systems cannot guarantee all of the required conditioning demand.

1.2 Issues Analysis in HVAC Water System

The HVAC system in large public buildings has numerous operating issues, and there is large energy conservation space. Among them, the HVAC water system (including cooling source and water transmission system) presents more common and prominent problems. Considering that the HVAC water system is the primary side system of the air-conditioning system, its operation characteristics are complex. And the energy consumption of the HVAC water system accounts for about 80% of the total energy consumption of the HVAC system. It is worth noting that although the efficiency of the chiller has almost doubled in the past 30 years, the proportion of energy consumption of HVAC water system still shows an increasing trend in recent years. Therefore, it is very necessary to conduct an in-depth analysis of the operating issues of HVAC water system in large public buildings. The main issues are as follows:

(1) The system configuration of HVAC water system is redundant, that is, the selected capacity of a single component and the designed component number are large. For the cooling source, on the one hand, the designed cooling load of system is rather large, which increases the reference value for determining the designed cooling capacity and number of cooling source. On the other hand, most of the cooling source selection does not take into account the dynamic load variation characteristics of each end in the system and the usage coefficient of building functions, resulting in a further increase in the design capacity of cooling source.

    For the pump, on the one hand, the selection calculation of the pump is usually too conservative, the design circulating resistance is too large, what is more, the circulating resistance is determined by estimation method during the construction drawing design step, rendering that the calculated circulating resistance is one or several times higher than the actual level, and the design head of the water pump is seriously higher than the actual system required level. On the other hand, the number of pumps in most systems is selected according to the principle of "one chiller for one set of chilled water pump and cooling water pump", which makes the excess configuration issues of cooling source extended to the water pump.

    Since the single component with rated operating condition deviates greatly from the actual system required operating level, the operating efficiency of the component cannot reach the ideal level, and the practical operating power consumption is naturally high. Taking the pump as an example, the mismatch between the pump performance and the water pipe network characteristics would make the pump work at a low-efficiency point for a long time. The measured results show that the maximum efficiency of some pumps can reach up to 70%, but the operating efficiency is only 30–40% in practice. The redundant configuration of the component number not only increases the initial investment of the system, but also causes resource wasting for long-term idle component.

(2) The operating issues of the transmission and distribution system are the root cause of the high energy consumption of the HVAC water system. The transmission and distribution system of the HVAC water system includes chilled water pump, cooling water pump, chilled water pipe network, and cooling water pipe network. Set the chilled water system as an example.

    Firstly, the hydraulic balance design of chilled water pipe network (user side) in most cases does not really meet the requirements. On the one hand, the chilled water flow of some branches in unfavorable loops still fails to meet user requirements even the system total chilled water flow is large. On the other hand, for those branch users with sufficient flow supply, the excess chilled water flow would make the regulating valve work at a low valve opening level for a long time, thus consuming a large amount of energy consumption. According to the literature, the regulating valve in the HVAC water system consumes about 33–50% of the power consumption of the pump. However, such a large cost does not bring back a good regulating effect.

Secondly, the operating state of the chilled water pump has not been effectively adjusted, so that the actual output of the transmission and distribution system is higher than the required level. Even setting aside the problem of low pump efficiency caused by redundant configuration, the operating state of the chilled water pump is still not ideal in practice. This is because the HVAC system operates in partial load conditions at most of the time, if the water flow provided by the transmission system does not adapt to the load variation and has maintained the water flow for design demand load, it is inevitable to make the water system operating in the condition of "large flow and small temperature difference". On the one hand, the pump would keep consuming energy that should be save. On the other hand, rendering that the temperature difference of the chilled water in the system under current operating condition is less than the reasonable level. The operating state with small temperature difference will lead to the low evaporating temperature of the chiller and reduce the operating efficiency of the cooling source. Since the energy consumption of cooling source accounts for the vast majority of the energy consumption of HVAC water system, this "small temperature difference" condition will lead to higher energy consumption of HVAC water system.

It needs to be pointed out that on the premise of a certain system load demand, there must be a "large flow" issue, and then, there will be a "small temperature difference" issue. That is to say, the unreasonable operating state of the chilled pump is the root cause of the high system energy consumption. The above analysis gives a reasonable explanation for why the energy consumption of the HVAC water system is still high when the performance of the chiller is constantly improving. The operation and regulation of the chilled water pump leads to the low-efficiency operation of the chiller for a long time, resulting in a substantial increase in system total energy consumption.

(3) The representative optimization control technique of variable water flow for HVAC water system is not applied widely or effectively. In order to solve the issues of (1) and (2) that objectively existed in HVAC water system, relevant optimization control methods and techniques are generally appeared. Among them, variable water flow technique is one of the most representative optimization control methods for the transportation system with the most serious issues. But in practice, there are few successful applications of variable water flow technique in China, and the main reasons can be attributed to the following two aspects:

Firstly, the main process of variable flow technique is to adjust the speed of chilled water pump or cooling water pump according to the load variation and to achieve the control of chilled water flow or cooling water flow. Commonly, the differential pressure of pump is often selected for reflecting the system load variation, that is, the change of the pump speed refers to the pipeline differential pressure. In which, how to select the proper differential pressure setpoints that can match the pump operating status and the demand energy requirements of end users reasonably is still an issue should be further studied.

Secondly, the constant differential pressure resetting strategy is widely used in the application of variable flow technique. However, if the differential pressure setpoint is too large, the strategy cannot fully stimulate the pump energy saving under partial load operating condition. If the differential pressure setpoint is too small, the strategy may not be able to meet the demand energy requirements of end users in some operating conditions. Therefore, the critical issue for the effective application of variable flow technique is how to take into account both the energy requirements of end user and the energy conservation effect of pump during the process of optimization of differential pressure.

1.3 The Decoupling Optimization Issues in HVAC Water System

The HVAC water system mainly includes the large mechanical and electrical components such as chiller, chilled water pump, cooling water pump, and cooling tower, as well as auxiliary components such as pipe network, end coils, and regulating valves. Figure 1.1 is a partial monitoring and control schematic diagram of a HVAC water system. As can be seen from the diagram, the various components in the system are closely related and interacted with each other. The monitoring and control points are numerous and spread throughout the whole building space, making it difficult to optimize and control the system. For example, when the end valve opening is regulated, the pipe network impedance will be changed and then affect the efficiency of the chilled water pump. And when the pump is regulated based on the differential pressure of pipe network, the chilled water flow will change, which will not only affect the chiller operating performance and the supply water temperature control effect, but also affect the control effect of end coils. Therefore, for the optimization and control issues of such complex system, it should not simply pay attention to the control effect of a single component or single control loop, but should pay attention to the comprehensive and overall performance of the system by considering the interaction among various components. In order to ensure that all components can operate cooperatively and efficiently, it is necessary to consider the correlation among all components and the coupling characteristics of parameters, as well as the parameters dynamic delay response characteristics in complex pipe network. In addition, the HVAC water system shown in Fig. 1.1 only depicts the pipelines, regulating valves and energy consumers on one branch of the pipe network. However, the pipe network and monitoring points of the actual system will be more complex and diverse, and the system will also show more complex nonlinear, strong coupling, and lag characteristics, which brings huge challenges to the decoupling optimization control of the HVAC water system.

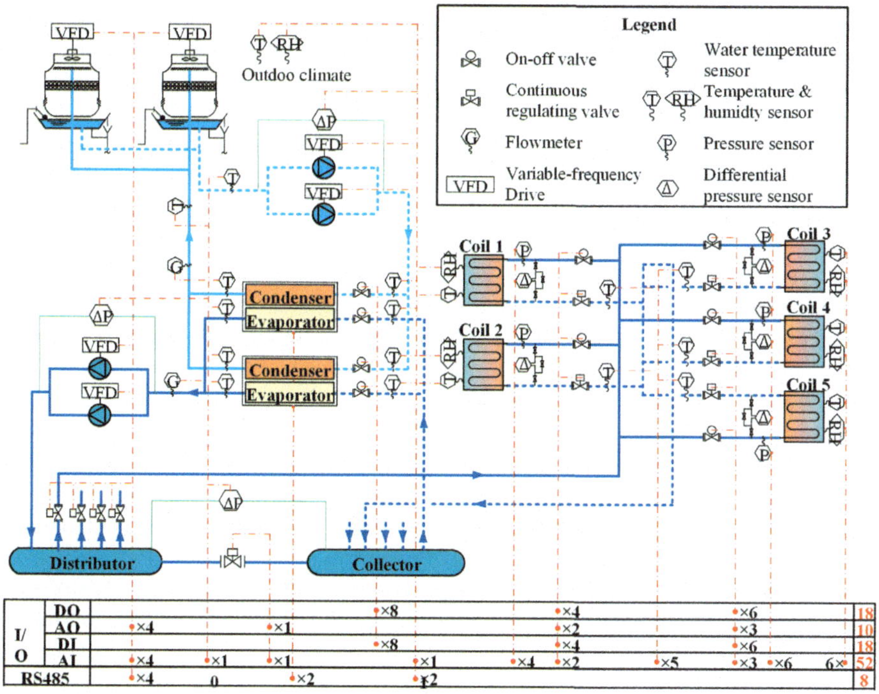

Fig. 1.1 Monitoring and control schematic diagram of general HVAC water system

In recent years, many scholars have conducted in-depth studies on the energy-saving optimization of each component and subsystem of HVAC water system, which has made great contributions to the efficient operation of cooling tower, cooler, chilled water pump, cooling water pump, and end coils. However, with the development tide of artificial intelligence and Internet of Things technology, users' demand for intelligent regulation and control of HVAC system is constantly increasing, and there are still some issues remained to be solved in decoupling control of HVAC water system. How to study the universal and efficient intelligent optimization control method according to the stochastic boundary conditions and complex nonlinear characteristics of the controlled object is the key to improve the global optimization control effect of HVAC water system, further improve the intelligent level and energy efficiency of buildings, and then create a healthy and comfortable indoor environment.

According to the difference of regulation parameters, the traditional regulation method for HVAC water system can be classified into quality regulation and quantity regulation. The parameters related to quality regulation can be collectively referred to as quality parameters, such as temperature and temperature difference, while parameters related to quantity regulation are collectively referred to as quantity parameters, such as flow rate and pressure difference. Among them, the dynamic propagation of the quality parameter in the pipe network shows slow time-varying characteristics, while the quantity parameter shows fast time-varying characteristics. With the development of equipment intelligence and automation technology, the control method of HVAC water system in actual operating process cannot be simply classified into quality regulation or quantity regulation, but the form of mixed regulation of quality and quantity regulation. Due to the difference in the time-delay response characteristics between quality and quantity parameters, the two kinds of parameter show significant coupling characteristics in the actual control process, which seriously affects the control effect of each control loop. Take the control loops in the cooling water system as an example, including the cooling water temperature difference control loop (loop 1) and the cooling tower outlet temperature control loop (loop 2). Among them, loop 1 is to control the temperature difference by adjusting the speed of the cooling water pump to change the flow of cooling water (quantity parameter), and loop 2 is to control the cooling tower outlet temperature (quality parameter) by adjusting the speed of the cooling tower fan to change the cooling tower air volume. When the temperature difference of cooling water is regulated, the water flow (quantity parameter) will change, which will affect the control effect of the cooling tower outlet water temperature (quality parameter). Similarly, when the cooling tower outlet water temperature is regulated, the cooling tower air volume (quantity parameter) will change, thus affecting the control effect of cooling water temperature difference (quality parameter). Similar coupling phenomena also exist in chilled water system and end coil system, and with the increase of system size and complexity, the coupling characteristics of the system will be more prominent, and the impact on the system operating performance and energy efficiency will be more severe. Therefore, the decoupling optimization control for HVAC water system by combining with the time-delay response characteristics of parameters and coupling mechanism is the key to improve the global operating energy efficiency of the system.

To sum up, the parameters like temperature and water flow have significant delay characteristics, the structure of pipe network is complex, and the control loops are coupled each other in HVAC water system, which shows significant nonlinear and coupling characteristics. In the research on the decoupling optimization control of HVAC water system, it is hard to guarantee the overall performance of the system only by considering the operating energy efficiency of one single component or single subsystem. How to deeply study the time-delay response characteristics of the parameters in the pipe network of HVAC water system by considering the nonlinear characteristics and exploring the coupling relationship among the control loops is the theoretical basis for realizing the decoupling optimization of HVAC water system and the key technique for improving the system overall energy efficiency.

1.4 The Optimal Issues of Control Period and Setpoint

At present, the optimization period and the adjustment value of setpoint for each control loop lack of scientific basis during the process of global optimization control of HVAC water system. The complex nonlinear characteristics of HVAC water system bring great challenges to the scientific resetting of the setpoint of each control loop.

The control rules of the controlled object with multi-time delay characteristics are difficult to be described based on the physical characteristics, which makes it hard to take into account the requirements of thermal environment regulation, energy conservation and thermal stability by fixed setpoints, and hard to grasp the optimal timing with fixed optimization period.

Although many scholars have conducted elaborate researches on the variable setpoint issues of HVAC water system, but most of them failed to coordinate the relationship between the optimization cycle and the set point adjustment amount, and there is a lack of linkage between these two terms. For a complex HVAC water system, the system required setpoint adjustment amount at different time stamps will be different, and how to determine the appropriate setpoint adjustment amount at the optimal timing, during the process of system optimization and regulation, is the key to improve the overall optimization control effect of the system. There are two key issues involved during this process. The first is how to determine the optimal timing, and the second is how to obtain the optimal setpoint at the right timing. Taking the chilled water system as an example, the variable flow control method is one of the most widely used energy conservation techniques in practical system. The control objective of this method is to reduce the energy consumption of the chilled water pump as far as possible on the premise of meeting the end load requirements and the component operating safety. The regulating basis is generally to adjust the pump speed to change the chilled water flow according to the differential pressure between the water supply and return of end coils. However, when the water flow changes, the control effect of chiller supply water temperature will be affected. How to select a reasonable differential pressure setpoint and supply temperature setpoint, so as to minimize the total energy consumption of related components on the premise of meeting the load demand, is still an issue worth studying. In existing studies, although some scholars have proposed some targeted control strategies, such as optimization strategies based on maximum valve opening and the most unfavorable thermal loop, all of them have not considered the coupling relationship among control loops, and also lack of more in-depth discussion and research. Some issues, such as unclear reference objects of optimization control and non-specific dynamic control schemes, have not been solved. Similarly, the issue that how to obtain a reasonable water temperature setpoint (such as chilled water supply temperature and cooling water return temperature) in the study of variable water temperature control method and how to obtain a reasonable valve opening setpoint in the study of the control method of end coil regulating valve are must be more in-depth studied by combining the coupling characteristics of HVAC system.

In addition, most existing studies use the fixed time intervals to optimize the system periodically. The size of the time interval is usually determined artificially according to the system size and operating experience, and the interval is generally selected between 15 min and 2 h. However, the control loops of the actual system will not be regulated frequently by considering the service life of the equipment, which makes the system under the fixed control period strategy hard to adapt to the complex and changeable operating conditions (such as outdoor weather fluctuations and large-scale personnel turnover), and the system is unable to respond in time and miss the best regulation opportunity. Moreover, when the system operates steadily and does not need to be regulated, the fixed period optimization strategy will still optimize the setpoint with the periodical time interval, which causes unnecessary optimization actions, further affecting the system stability. Therefore, how to study the intelligent control method with variable control period and variable setpoint on the basis of the intelligent control requirements, so as to properly regulate the system at the right timing, is the key technique to realize the decoupling optimization of HVAC water system and improve the system operating intelligent control level.

1.5 Summary

The HVAC water system displays significant delay characteristics in parameters such as temperature and water flow, and its pipe network structure is intricate. Furthermore, the control loops are interconnected, leading to pronounced nonlinear and coupling features. When striving for decoupling optimization control in HVAC water systems, relying solely on the consideration of the operational energy efficiency of individual components or subsystems presents challenges in ensuring overall system performance. The theoretical basis for achieving decoupling optimization lies in a comprehensive examination of the time-delay response characteristics of parameters within the pipe network of the HVAC water system. More studies should encompass nonlinear characteristics and delve into the coupling relationships among the control loops. Such an approach is deemed a vital technique for enhancing the overall energy efficiency of the HVAC water system.

Meanwhile, there are significant challenges in globally optimizing HVAC water systems, particularly due to the absence of a scientific framework for establishing the optimal period and adjustment values for setpoints in each control loop. The intricate nonlinear characteristics of HVAC water systems present substantial obstacles when it comes to scientifically readjusting the setpoints of control loops. The controlled entities, characterized by multi-time delays, complicate the formulation of control rules based on physical attributes, thereby hindering the management of thermal environment regulation, energy conservation, and thermal stability through fixed setpoints. Moreover, there is difficulty in pinpointing the ideal timing within a fixed optimization period. In essence, the lack of a scientific underpinning for determining setpoints and optimization periods emerges as a critical concern in the current global optimization of HVAC water systems.

Chapter 2
State of the Art

2.1 Modeling Methods of HVAC System

To ensure proper functioning, minimize environmental impact, and provide optimal indoor conditions for occupants, HVAC systems must be designed, operated, and maintained efficiently. Optimization and control methods are the most commonly used energy conservation techniques in HVAC systems. Modeling HVAC systems is an essential step toward conserving energy and improving indoor air quality, prior to implementing optimization and control. Currently, the most common methods for modeling HVAC systems can be classified into three types: data-driven modeling, physics-based modeling, and gray box modeling; each method has its own unique advantages and disadvantages.

The data-driven modeling approach employs mathematical techniques to establish a correlation between input and output variables by utilizing historical operating data collected under normal usage or specific test conditions. This category of modeling methods includes artificial neural network (ANN) and statistical regression modeling approaches. Despite their high accuracy, data-driven models are limited in their universality beyond the training domain and are only suitable for modeling systems or components that retain long-term historical data.

In contrast, the physics-based modeling method employs the laws of physics and system/component knowledge to reveal underlying processes, resulting in detailed mechanisms of analyzed systems. The derived models are also known as white box or forward models. While requiring significant effort for development and calibration, physics-based models possess strong universality capabilities. Once developed, these models can be converted to frequency domain transfer functions or time-domain state-space representations, providing a valuable tool for analyzing and modeling HVAC components. Despite their complexity, the major applications of physics-based models in HVAC components make them a useful tool for engineers seeking to gain greater knowledge and understanding of these systems. However, their inherent complexity may lead to poor accuracy.

T. Zhao et al., *Thermal Distribution Principle and Nonlinear Optimal Control for HVAC Water System*, Indoor Environment and Sustainable Building,
https://doi.org/10.1007/978-981-96-0377-0_2

For the gray box modeling method, it aims to combine the advantages of both data-driven and physics-based modeling method by using the structure of physics-based models but estimating model parameters using system performance data. This renders the models inherit the advantages of physics-based models and data-driven models and present good universality capabilities and better accuracy.

In conclusion, each modeling approach has its own set of advantages and disadvantages. While data-driven models offer high accuracy on training data, they suffer from poor universality. On the other hand, physics-based models provide good universality capabilities but sacrifice accuracy. Gray box models combine the strengths of both approaches to provide good universality and accuracy. Therefore, the selection of modeling approach is ultimately contingent upon specific application requirements and available resources. Herein, the literature review of modeling methods for HVAC systems will be presented from three perspectives: component-level modeling methods, coupling characteristic modeling methods, and system control-oriented modeling methods.

2.1.1 Modeling Methods of Components

In HVAC systems, primary and secondary components work together to provide heating and cooling to buildings. Primary systems utilize fuel and electricity to generate thermal energy, which is then transferred through secondary systems for distribution within the building. Examples of primary components include chillers, boilers, cooling towers, and thermal storage systems. Secondary systems consist of air handling units, air distribution networks, and chilled water distribution networks that connect the primary system with the interior of the building. Both primary and secondary systems comprise distribution components, such as pumps, fans, dampers, valves, ducts, and pipes. These components impact the energy flow in buildings by consuming electrical energy to drive pumps and fans and by transferring thermal energy to and from the fluid in all distribution components [1].

The models utilized for describing distribution components must adhere to energy and mass balance equations, yet many building performance simulation tools over-simplify these models [1], thereby eliminating the necessity of calculating pressure drops through distribution systems during off-design conditions. While this approach suffices for examining thermal properties of system, a thorough analysis of fan and pump control loops, as well as issues regarding the placement of return (or exhaust) fans and the type and size of dampers (or pipes), necessitates consideration of flow and pressure balancing between components [2].

The HVAC system comprises electrical components (such as chillers, pumps, cooling towers, coils, and air handling units) and accessory components (such as pipe/duct networks, valves, and dampers). The majority of component models utilized for building and system performance simulations exhibit the following characteristics [3]:

a. They exhibit continuity in state, implying that the range of model variables is represented by real numbers or intervals. However, certain models may assume a discrete set of values, thereby rendering them as discrete state models.
b. They are discrete in time for better application. If the models demonstrate continuity in state and discreteness in time, they are typically described by a (set of) difference equation(s).
c. They are deterministic, although stochastic models are also used in certain applications such as predictive control [4].
d. The rules of interaction change over time, making these models time-varying.
e. They can describe both steady state and dynamic conditions.
f. They are mainly used for forward prediction, meaning that they predict the response of output variables based on a known structure and known parameters when subjected to input and forcing variables.

For the modeling of HVAC components, most researchers prefer to the physics-based modeling method to build the model for the components with clear mechanism. For example, the requirement of zone model is to predict the indoor temperature and heat transfer process. The underlying mechanism of zone model follows the energy conservation law [5], such as the heat conduction of building envelope [6]. Therefore, the model can be derived from heat conduction and radiation rules by using the energy balance laws and transfer function method [7, 8]. Besides, by considering the exhale rate of CO_2 and humidity of occupants, the predicted model of indoor CO_2 and relative humidity also can be derived [9, 10]. Similarly, the thermal radiation model between two surfaces with a large temperature difference follows the Stefan–Boltzmann law, and it must be considered when the surface radiates thermal energy to the zone for an extended period of time [11]. For the cooling (or heating) coil, the heat exchange on water side is identical to the air side, and the model also can be derived by the energy balance laws [12, 13]. Furthermore, the physics-based modeling method is also appropriate for the components with mechanical characteristics, such as the valves [12], fan and pumps [14–16], and compressor [17]. The Merkel's method is often used to build cooling tower models [18, 19], which follows the energy balance laws and is also a kind of physics-based modeling method.

As the development of data-driven modeling method, it is becoming more acceptable for the component models, and the models are able to present the real characteristics of component. For example, the transfer function coefficients of air temperature in AHU can be identified by the model-fitting techniques [20–23]. For the complicated or nonlinear characteristics, the data mining techniques are often applied to the modeling of system components, such as the artificial neural network (ANN) models for fault detection of cooling coil [24], the support vector machine (SVM) models for chiller cooling load prediction [25], and the statistical model (autoregressive moving average model) for indoor cooling load prediction [26]. Some hybrid method, like particle swarm optimization algorithm based SVM (PSO-SVM) model, is also proposed for predicting system cooling load [27], in which the PSO algorithm is used to calculate the coefficients of SVM model. The PSO-SVM modeling method is able to overcome the premature problem and output the convincible parameters.

2.1.2 Modeling Methods of Coupling Characteristics

The control approach for a pipeline or duct network system can be divided into two categories: quality regulation and quantity regulation [28, 29]. The quality parameters, such as water and air supply temperature and temperature difference, collectively control the quality regulation, while quantity regulation is controlled by the quantity parameters, such as water and air flow rate and differential pressure. The coupling relationships between control loops in variable flow rate HVAC systems are largely attributed to the dynamic diffusion characteristics of quality and quantity parameters in the pipeline network [30]. These dynamic diffusion characteristics reflect the gradual process of parameter response along the pipeline, from the source (such as a chiller or air handling unit) to the end (such as a coil or variable air volume box), as control actions are implemented. When a new setpoint is introduced into the system, for example, the temperature or flow rate adjusts gradually along the pipeline until it reaches the desired level.

Aiming at the cooling water system of conventional central air-conditioning system, Lu [31] considered the heat exchange mechanism to build the coupling characteristics models of cooling tower, studied the influence of cooling tower air volume and cooling water flow on the global power consumption of the whole system, the optimal cooling tower air and cooling water flow can be obtained by adopting genetic algorithm. Liu [32] established a series of HVAC system models, including conditioned zone model, component model, and pipe network model. The key parameters between related models are specified to describe the coupling characteristics among components, and the interior point method was used to solve the nonlinear convex optimization problems. Zhuang [33] has divided the entire HVAC water system into three subsystems according to the spatial distribution characteristics: cooling tower system, chillers' system, and end coil system. The coupling characteristics in system were described with interaction between subsystems. Inspired by the sequence optimization, a distributed control method has been proposed to obtain the optimal solution, which makes the optimization objectives converge through the information interaction and mutual influence among the subsystems.

The interdependence of control loops is primarily expressed by the relationship between temperature and flow rate during the control process. Specifically, controlling the flow rate can cause temperature fluctuations, while controlling temperature can affect the required flow rate. This unpredictability can have negative impacts on energy efficiency and overall system control. Many techniques have been developed to handle the unpredictability, but they often require complex mechanisms and may not be easily implemented in practical scenarios, such as the techniques of feedforward decoupling network [34, 35], differential pressure reset [36, 37], and relative gain array [38, 39]. Some studies address coupling relationships for optimization, but they often lack universal applicability, such as the research on distributed architecture [40], novel system configuration [41, 42], and model predictive control [43–45].

Theoretically speaking, the response time difference between quality and quantity parameters can exacerbate coupling relationships during the control process [46, 47], particularly when setting optimal setpoints. Any changes in the setpoints will cause dynamic parameter diffusion in the pipeline or duct network, influencing the thermal and hydraulic stability at end users. Current research on temperature resetting [48–50] or differential pressure [51–53] rarely addresses this issue systematically, primarily discussing delayed response impacts. Therefore, in the optimization process of HVAC system, it is essential to consider the basic decoupling mechanism when studying variable flow rate HVAC systems.

2.1.3 Modeling Methods for System Control

There are two types of HVAC controllers based on different control tasks, including the local controllers and supervisory controllers [54].

The local controllers ensure proper HVAC system operation and service delivery. They are mainly oriented to the control of specific equipment and can be further classified into sequencing controllers and process controllers [55]. Where, sequencing controllers determine the order and conditions for switching equipment on and off, including chiller, cooling tower, pump, and fan sequencing controllers. Process controllers regulate control variables to maintain the desired setpoint in the case of existed disturbances and system dynamics, and common process controllers in the HVAC industry include proportional (P), proportional–integral (PI), proportional, integral, and derivative (PID), ON/OFF, and step controllers.

Supervisory controllers are advanced controllers that take into account the system-level characteristics and interactions between all components and their associated variables. They are mainly oriented to the system control loops and responsible for optimizing operation modes and setpoints for local controllers, such as the application in cooling water systems [56–58].

From the perspective of application, controllers should be expressed by equations in discrete form. They can facilitate the interaction between the building and the system, as well as the interaction between components within the system. In practice, a closed-loop control involves a sensor that samples a measurable variable from the field. The controller calculates the control signal based on the setpoint and the measured value by using a control algorithm and outputs a specific regulating quantity; this signal is then sent to the real-world actuator to reach the control objective.

Recently, with the high integration density of simulation tools, the modeling method for system control has been integrated in these tools, such as EnergyPlus [59–61], TRNSYS [62], and MATLAB [63, 64]. In which, users can directly select the proper control strategy and set corresponding parameters. However, simulation tools allow users to manipulate variables that cannot be sensed or actuated in reality, as well as apply control algorithms that do not exist in reality. For instance, a modeler can directly control the heat flux in a simulation, whereas in reality, this can only be done indirectly by changing a valve or damper position.

Moreover, simulation allows for the possibility of achieving ideal control, which is not always practical in the real world due to the inaccessibility of certain variables such as zone cooling/heating load. An ideal local-process controller can adjust the actuated variable to meet the setpoint requirements without specifying a specific control algorithm, by numerically inverting forward simulation models to calculate the necessary input to achieve the desired output.

The capacity to simulate diverse advanced controllers using modern tools has certain restriction. Certain tools provide predetermined control strategies (system-based simulation tools), while others offer flexibility in specifying only supervisory controllers [59], or ON/OFF controllers [60], or even local controllers [62]. Although domain-independent environments like MATLAB and Dymola are effective for designing and testing controllers in a simulation environment, they lack models for all other physical components and phenomena in buildings.

2.2 Optimal Control Methods in HVAC System

2.2.1 Conventional Control Methods

The conventional control methods use classical control strategies and have been in use for many decades, including feedback control and proportional, integral, and derivative (PID) control. These classic models also include manual control, feedback control, feedforward control, On–Off, proportional (P), proportional–integral (PI), among others. The simplest control technique within this category is the bang–bang control, which employs a thermostat to provide temperature feedback. In the cold weather, a heating unit is activated when the temperature drops below a predetermined threshold until it reaches the setpoint, while in the hot weather, a cooling unit is activated when the temperature exceeds a specified point until the temperature returns to the setpoint. Despite being a fundamental and easy-to-implement control method, this method is highly energy inefficient and unable to precisely track the setpoint due to frequent temperature overshoots [65].

PID control is commonly used to address various issues in building automation. While On/Off or bang–bang control is still used in older structures without digital control, more contemporary buildings with variable frequency drivers often utilize PID control instead. Liu [66] provided an example of PID control through the construction of a portable device for automated tuning of PID regulators in HVAC system. In Sundriyal's research [67], a classical control method based on PID controllers is discussed, with the authors developing a thermal model using a first-principle method and using transfer functions to simulate PID controllers in heating and cooling modes.

While PID controllers are easy to build and relatively inexpensive, the parameters can be laborious to tune and may not address energy efficiency issues or account for real-time restrictions and nonlinear dynamics. To address these issues in HVAC systems, rule-based control methods are often employed. These controllers use a set of "if–then–else" rules written by skilled property managers to operate each HVAC component independently and adjust the setpoints of each PID loop based on specified control rules. The rules are often determined by the experience of manager. Saloux [68] developed a control methodology based on rules to regulate the water flow of a heating system equipped with energy storage, and the method led to a significant 43% reduction in operational costs. Similarly, Ali [69] proposed a rule-based control scheme for a heat pump system combined with a thermal energy storage tank, which resulted in a 15% decrease in operational costs. These studies suggest that rule-based control is a feasible and effective approach for the HVAC systems involving thermal storage. However, it's worth noting that rule-based techniques are typically employed to regulate the specific component without optimization at the system level. Although local controllers may be complex, the absence of upper layer controller that supports the optimization of setpoints is still a challenge. This limitation can be attributed to the complexity of upper layer requirements and the difficulty in generalizing their regulations at the system level.

Another conventional control method is the fuzzy logic control, which has the potential to address all HVAC system requirements and components by integrating them [70]. In an article cited as [71], the use of fuzzy logic in HVAC control is demonstrated. The authors utilized a fuzzy controller to manage HVAC units in four rooms of a building where there was insufficient energy to keep them all on. The fuzzy controller ensured that internal thermal comfort was maintained while preventing situations where the HVAC power demand exceeded the available power, thus achieving peak-load reduction. The authors successfully demonstrated their controller in a laboratory environment. This approach was similarly employed in [72] where two fuzzy controllers were used. The first controller was used to analyze energy consumption and availability, while the second controller allocated the available energy to maintain a comfortable temperature in each zone.

2.2.2 *Variable Setpoint Control Methods*

For the optimal control methods of HVAC water system, the variable setpoint control is the most used technique in practice. From the optimization control process of the HVAC water system, the first step is the setpoint optimization of each control loop, that is, under the premise of ensuring the operation requirements of the system and the safety of the equipment operation, through the reasonable resetting of the component operation parameters so as to achieve the high efficiency and energy saving operation goal. Su [51] proposed an optimization method for determining the optimal pressure difference setting value of chilled water pipe network by applying Hazen–Williams formula, and the engineering application results showed that this method could effectively reduce the system energy consumption. Zhao [73] proposed an online optimization method based on the most unfavorable thermal loop for resetting of refrigerant water pressure difference. Engineering application results show that this method has high stability and may save 22–53% of pump energy consumption compared with the traditional pump control strategy. And then, Sun [36] studied the general calculation method for the resistance coefficient of water pipe network required for the calculation of differential pressure setting value in combination with the pump operating point and the resistance characteristics of water system, which is of great significance for the process of resetting differential pressure setting value. The above studies are all aimed at setting optimization of a single control loop. Reasonable resetting of flow and temperature setting of the control loop can effectively improve the operational energy efficiency of all kinds of equipment. Although the above studies have given a more effective optimization scheme for variable setting of HVAC water system, most studies focus more on the optimization of the set point of a single control loop and do not consider the coupling characteristics of temperature and flow between different control loops or different devices triggered by variable setting control. From the perspective of global optimization of the system, these optimization schemes are not the most ideal choice.

After determining the setpoint of control loops, it is necessary to optimize the state parameters (such as ON/OFF state, and speed) of parallel components and rationally distribute the given setpoint among parallel components, so as to minimize the overall energy consumption of same kind components on the premise of meeting the requirements of the setpoint. Therefore, the distribution optimization of similar equipment also plays a very important role in the optimization control of the global variable setting of the HVAC water system. The distribution optimization of HVAC water system includes cooling capacity distribution of chiller, head distribution of water pump, and heat dissipation distribution of cooling tower.

For the purpose of simple operation and easy maintenance, chiller units with the same capacity are generally selected in the multi-unit system design [74]. However, some studies have shown that the combination of chiller units with different capacities can more flexibly adapt to changes in building load [75]. In 2004, Chang [76] proposed a load distribution optimization method based on Lagrange method for a system with multiple chiller units in parallel. This method takes the COP sum of the chiller as the optimization objective and the total cooling capacity of the chiller equal to the cooling load of the system as the constraint condition, which can effectively reduce the complexity of the nonlinear equation. It has high feasibility and good energy saving potential. In the following year, the author [77] adopted genetic algorithm to improve the previous Lagrange method, and the improved method has higher computational accuracy. In view of the disadvantages of large calculation amount of traditional optimization methods, Chang [78] proposed that dynamic programming method could be used to optimize the start-stop sequence of chiller, so as to make up for the defects of traditional methods. Experimental results showed that this method could achieve optimal load distribution of chiller with a lower calculation amount. The above research shows that the load distribution optimization of the chiller unit can be achieved by using the traditional optimization method through the modeling of the system. However, such method has a high dependence on the model and needs more accurate modeling support.

For the study on distribution optimization of water pump, some scholars have discovered certain operation rules and proposed corresponding control ideas and rules based on the physical characteristics of water pump. For example, based on parallel hydraulic characteristics of variable frequency water pump and extreme value analysis method, Zhao [79] proposed in 2012 to study the optimal operation combination and speed ratio allocation of parallel water pump. Wu [80] divided the pump system into four parts according to its dynamic characteristics, which were optimized by artificial neural network to improve modeling and computational efficiency. In addition, some scholars have also used swarm intelligence algorithm to study the distribution optimization problem of water pump, and the research results show that swarm intelligence algorithm has higher efficiency and stability than traditional algorithm [81, 82].

For the research on distribution optimization of cooling tower, although the energy consumption of cooling tower is very small compared with that of water chiller and water pump, because it undertakes the heat dissipation task of the whole system, its heat dissipation capacity will indirectly affect the energy consumption of other two, so the research on cooling tower should be based on the premise of meeting the heat dissipation. In practical applications, the cooling amplitude of the cooling tower is usually used as the basis for the control of the fan speed ratio, that is, the outdoor wet bulb temperature is taken as a reference to control the water temperature of the cooling tower, so that the difference between the two is always kept within the optimal cooling amplitude range. In terms of the control strategy of the number of cooling towers, the operation strategy of "one machine, one tower, one pump" is a control method commonly used at present, that is, the start-stop setting of cooling towers is the same as that of corresponding chillers. In addition, another common control scheme is "uniform water distribution and synchronous frequency conversion". Under this strategy, the number of cooling towers should be opened as much as possible under the premise of meeting the lower limit of the water flow of the cooling towers, the cooling water is evenly distributed in the open cooling towers, and the fans of each cooling tower perform synchronous frequency conversion according to the water outlet temperature. Compared with the traditional control strategy of "one machine, one tower and one pump", this control mode can make more full use of the heat transfer area of the cooling tower, reduce the resistance of the cooling water side, and effectively improve the heat transfer efficiency of the cooling tower [83].

In summary, research on variable setpoint control methods for HVAC water systems has primarily focused on the subsystem optimization control, with limited attention given to system-wide analysis and spatial characteristics. And the optimization control algorithms for optimal setpoints have prioritized improving accuracy, with less consideration for practicality and online feasibility. While some algorithms have shown promising experimental results, they are often specific to certain systems and difficult to apply in practical engineering. Given the spatial correlation and coupling relationship among control loops, it is crucial to analyze the system's spatial characteristics and conduct comprehensive research on the system as a whole while optimizing control loop setpoint.

2.2.3 Intelligent Control Method

In the 1980s, there was a growing focus on the development and implementation of intelligent methods for building control systems [84]. The intelligent controllers could be customized to manage the numerous subsystems in modern buildings by utilizing the evolutionary algorithms. The artificial intelligence (AI) derived learning systems offers a unique way for addressing energy management concerns, which differs from traditional methods. AI-based controls can accommodate partial operating conditions or other noisy inputs and are capable to handle system nonlinear characteristics. With proper training, AI controls can make predictions and regulate the HVAC system effectively [85]. However, with the restriction of hardware performance and implementation feasibility, the intelligent methods are not popularized at that time.

Recently, with the raising development of hardware performance and artificial intelligent techniques, the intelligent control methods for HVAC systems have been frequently mentioned in most researches and are expected to be the key techniques of the intellectualization of HVAC system. The most frequently employed AI technique is artificial neural networks (ANNs), which have been extensively utilized for predicting building and HVAC management techniques [86], such as the improved sensitivity analysis combination method based on ANN for optimizing setpoints in HVAC system [87]. The deep neural networks (DNNs), utilizing multiple hidden layers to effectively organize hidden neurons, are a dependable method for modeling intricate nonlinear relationships, surpassing traditional ANN in performance [88], such as a DNN-based mixed-integer linear programming method for HVAC system energy efficiency and thermal comfort [89]. These studies indicate that the neural network-based control method has better performance on handling system nonlinearity and random conditions.

In recent years, the data acquisition system has significantly advanced, making it more convenient and cost-effective to gather and retain building data. And the use of big data technology enables better optimization of HVAC systems. These new approaches, including the stochastic model predictive control (SMPC) method [90] and the deep reinforcement learning (RL) method [91–93], have become increasingly popular in HVAC systems for improving thermal comfort and energy efficiency primarily because they are based on real-time operational data, which enhances their accuracy and dependability [94]. Du [95] utilized a unique model-free deep RL method to minimize energy consumption in HVAC systems, with the findings indicating that the approach can reduce energy costs by 15% compared to rule-based control. Yang [96] proposed a gradient-based learning method for HVAC systems, and the results indicated a superior performance than model predictive control method on the terms of energy cost and handling uncertain thermal comfort.

Due to its parallel computing abilities, multi-agent-based control methods are gaining more attention recently in response to the high computing performance requirements of intelligent control methods. Zhuang [33] proposed a decentralized control method for a typical HVAC system to search for optimal setpoints.

Comparison with centralized control methods has shown that the proposed method reduces energy consumption by 34% and requires less computational time. Wang [97] proposed a multi agent-based control method with a hierarchical architecture that divides the overall optimization task into several subtasks by considering the interactions among components. The method reduces energy consumption by 4.05% and saves 69% of computational time compared to the conventional centralized control method. These studies indicate that the multi agent-based control methods offer unique strengths in computing efficiency, making them a promising solution for the intelligent development of HVAC systems.

2.3 Optimization Algorithms for HVAC Nonlinear Problems

2.3.1 Conventional Algorithms

Traditional optimization methods include linear programming, nonlinear programming, integer programming, dynamic programming, and duality theory. These methods have been widely used in various fields, such as industry, economics, management, and decision-making, to provide effective solutions for practical problems. Among them, the linear programming and nonlinear programming are the most commonly used methods for solving linear and nonlinear optimization problems, respectively. Integer programming is used for solving optimization problems involving discrete variables, such as the ON/OFF control of HVAC system [98, 99]. Dynamic programming is used for solving optimization problems with stage-wise and recursive properties, such as the optimal solution for cooling load distribution [100, 101]. Takabatake [102] adopted Bayesian optimization algorithm for searching the optimal setpoints of a typical HVAC system, and 10.4% of energy consumption can be saved. While these traditional optimization methods remain powerful tools for solving many practical problems, their limitations are becoming increasingly apparent with the rise of problem complexity and scale. Therefore, more advanced and efficient optimization methods are needed to meet these challenges.

2.3.2 Heuristic Algorithm

Heuristic algorithms are a type of algorithm that relies on experience and intuition to solve complex optimization problems. While these algorithms do not guarantee finding the global optimum, they can still find a solution that closely approximates the optimal solution within a reasonable amount of time [103]. Heuristic algorithms take inspiration from certain natural phenomena, such as evolution, swarm behavior, and simulated annealing. Examples of common heuristic algorithms include genetic algorithms (GA) [104], particle swarm optimization (PSO) [105], and equilibrium optimization (EO) [106]. These algorithms have been successful in solving complex optimization problems and have found widespread use in practical applications, including machine learning, data mining, and intelligent optimization. Heuristic algorithms offer several advantages, including the ability to handle high-dimensional, nonlinear, non-convex, non-smooth, and constrained optimization problems, as well as high levels of robustness and adaptability.

Researchers in HVAC system optimization have applied different algorithms to tackle the nonlinear and multi-objective problem [107, 108]. According to a literature review statistic [94], the GA is the most common algorithm used by researchers. Seong [109] used GA to calculate the optimal setpoints of HVAC system control loops, which can save 5.7% of energy consumption comparing with conventional method. In Arabali's research [110], a GA-based approach is employed to optimize a hybrid-renewable generation and energy storage system for meeting the HVAC system load. Results of simulations indicate that combining the GA with a two-point estimation can greatly increase the efficiency of the hybrid system. This leads to less energy waste and higher efficiency, as compared to classical optimization approaches. Ascione [111] uses a multi-objective GA (MOGA) to tackle a multi-criteria decision-making problem based on offline weather data. Simulation results provide practical insights for achieving considerable improvements in energy performance in Italian residential buildings. Authors suggest that the approach may be easily extended to other locations and different building types. In Papadopoulos's research [112], a MOGA-based simulation approach is proposed to adjust the heating and cooling temperature setpoints of HVAC systems in large office buildings. Authors conducted different experiments in various US climate zones to determine the link between location, climate conditions, and potential energy savings. Based on the results obtained, targeted policies and strategies are suggested to guide the creation of sustainable and user-centric built environments that include occupant interactions in the modeling process. Pezeshki [113] employed the GA to identify the optimal location for HVAC systems, with the aim of maximizing thermal comfort and minimizing energy consumption. The proposed GA-based approach endeavors to achieve the best results by tackling obstacles such as unknown fitness conditions, instability, noise, and local minimum. Simulations conducted at an Iranian hotel indicate that the GA-based approach outperforms similar state-of-the-art approaches. Besides, the methodology based on GA proposed in this article can be used globally in the construction industry by project developers, policymakers, and researchers as a novel approach.

Different algorithms have been used extensively to study optimization problems related to HVAC systems, specifically genetic algorithms (GAs) and swarm intelligence (SI)-based algorithms like PSO which have become increasingly popular [111–113]. GAs have high capability in solving complex optimization problems which involve large datasets and parameters that have nonstationary, nonlinear, discontinuous, and noisy objective functions. In contrast, PSO algorithms have better convergence rates and computation precision when compared to GAs. A novel method was proposed in [114] which is based on a comfort level indicator and a multi-objective scheduling model. The method was resolved by a MOPSO approach to achieve demand response in residential smart grids. The results of the experiments showed that the proposed method effectively optimized energy consumption while also enhancing users' comfort levels. Additionally, Yang [115] proposed a hierarchical HVAC control structure, which has an embedded PSO optimizer which is used to reduce energy consumption in an office building while maintaining comfort levels. Kusiak [116] proposed an ANN-based predictive model which was optimized using a strength- multi-objective-PSO algorithm. The model aims to optimize HVAC in office buildings while analyzing the relationship between energy consumption and thermal comfort. In some researches [105, 117], PSO optimization techniques were used to optimize PID controller parameters. The aim was to achieve better temperature control in HVAC systems. The results demonstrate the potential of PSO and other SI-based algorithms in optimizing both the internal control parameters of the model and solving the multi-objective HVAC optimization problem. This allows for minimizing energy consumption as well as maximizing user comfort levels.

The above studies show that compared with traditional optimization methods (Lagrange method, extremum search method [118], etc.), heuristic search algorithms (GA, PSO [119], etc.) have more significant advantages in solving complex optimization problems such as load distribution, and the algorithm has higher computational efficiency.

2.3.3 *Intelligent Algorithm*

Nowadays, with the development of computing performance, some more intelligent algorithms have been derived, such as the data mining algorithm [120], the decentralized swarm intelligence algorithm [121], and deep reinforcement learning algorithm [122]. Chen [123] proposed a fuzzy neural PSO algorithm to solve the optimal chiller loading problem for achieving the minimum energy consumption. Through comparison with linear regression method and equal load distribution method, the proposed algorithm can converge quickly in the process of solving the optimal loading problem. And high precision results can be obtained in a short time. Afroz [124] built the predictive models in HVAC system based on ANN and used the PSO algorithm to obtain the optimal setpoints, which can save 7.8% of energy consumption.

Some scholars use distributed swarm intelligent algorithm to solve the optimization problem HVAC system, and the results show that swarm intelligent algorithm has greater advantages in computing efficiency than traditional algorithms [125, 126]. In Wang's research [121], a decentralized swarm intelligence algorithm has been proposed, which was running in a distributed architecture. The findings indicate that the proposed algorithm has certain energy conservation potential and feasibility in hardware test. Liu [127] aimed at the optimal chiller loading problem in HVAC system and proposed a distributed chaotic estimation algorithm, in which the chaotic mutation mechanism is applied to increase the search capability of the algorithm.

By comparing with other conventional algorithms, the distributed algorithm exhibits better performance of robustness, stability, and convergence.

With the increasing popularity of machine learning methods, Mozer [128] firstly applied RL algorithms for the HVAC systems. He installed a smart control system in an existing building that can automatically adjust both HVAC and lighting systems according to the occupants' preferences. Recently, RL algorithms have been developed to optimize the HVAC systems, water heaters, energy storage units, etc. Biemann [129] compared four novel RL algorithms for continuous control, including soft actor critic, twin delayed deep deterministic policy gradients, trust region policy optimization, and proximal policy optimization, and the performance was evaluated on a common open-source environment in terms of energy efficiency, robustness, thermal stability. Several studies provide additional information on the use of RL in demand response applications. Henze [130] and Liu [131] employed RL to control thermal energy storage systems in the building energy sector. The results indicate that an RL algorithm can reduce the system costs, but requires a significant amount of data for training to achieve performance comparable to conventional predictive methods. Liu and Henze [132] proposed a technique of pre-training the agent in a simulated environment before deploying it in a real-world building, which is widely used today. The simulation model does not need to be completely accurate, but only needs to have the same states and actions as the learning controller during deployment. The same network and weights are used, but they must be regularly updated to improve performance and reduce the impact of the transition from simulation to reality. Liu [122] proposed a multi-step predictive deep RL algorithm for improving the energy efficiency of HVAC system, and the algorithm was able to flexibly adjust the power usage of HVAC system based on the changing electricity prices and can save more than 12% of energy cost. Huang [133] proposed a multi-step actor-critic algorithm for solving the energy conservation optimization problem of HVAC system. The algorithm employs a multi-step action-network and search tree to update the optimal state, that renders the strength on computation and convergence time.

The above researches indicate that intelligent algorithms are increasingly being applied in the field of HVAC system control. These algorithms offer superior performance and are better suited to meet the demands of today's intelligent control systems, making them a promising direction for algorithm development.

2.4 Summary

Through the analysis of the research progress outlined above, it is evident that the global decoupling optimization control of central air-conditioning systems presents a complex nonlinear optimization challenge. As research in this area continues to deepen, the integration of intelligent control methods such as fuzzy neural networks, genetic algorithms, and particle swarm optimization algorithms has introduced novel approaches to enhancing the decoupling and control efficacy of air-conditioning systems. Moving forward, several key issues persist in the realm of global decoupling optimization control for air-conditioning systems:

(1) The majority of coupling issues within HVAC systems stem from the interconnected parameters inherent to the system. While some scholars have proposed decoupling control and fuzzy control methods, these are primarily based on algorithmic simulations of decoupling and control within air-conditioning subsystems. There remains a dearth of research addressing the underlying propagation mechanism of parameters and the coupling dynamics of control loops at their core.
(2) In the domain of air-conditioning system optimization control, the predominant focus lies on "how to optimize", with scant attention paid to the temporal aspect of optimization. Given the multifaceted and dynamic operational conditions of air-conditioning systems, it is imperative to delve into the question of "when to optimize" should occur.
(3) Presently, the bulk of studies center on optimizing control of variable set values for local equipment within the system. However, from a holistic perspective, there exists a gap in research concerning optimization control methods for variable set values that account for the spatial correlations among various equipment and the coupling characteristics across control loops.
(4) Existing optimization control methodologies predominantly target specific systems, lacking in generalizability and reproducibility. Despite considerations of algorithmic computational efficiency and solution accuracy, there is a notable scarcity of research on universally applicable global optimization control methods tailored to air-conditioning systems of varying complexities.

References

1. Ashrae Handbook (2008) American Society of Heating, refrigerating and air-conditioning engineers. Int J Refrig 2:56–57. https://doi.org/10.1016/0140-7007(79)90114-2
2. ASHRAE Transactions (1998) A standard simulation test bed for the evaluation of control algorithms and strategies
3. Zeigler BP, Muzy A, Kofman E (2019) Theory of modeling and simulation. Elsevier. https://doi.org/10.1016/C2016-0-03987-6
4. Clarke JA, Cockroft J, Conner S, Hand JW, Kelly NJ, Moore R, O'Brien T, Strachan P (2002) Simulation-assisted control in building energy management systems. Energy Build 34:933–940. https://doi.org/10.1016/S0378-7788(02)00068-3
5. American Society of Heating, Refrigerating and Air-Conditioning Engineers, Inc. (2009) Energy estimating and modeling methods. In: ASHRAE handbook fundamentals
6. Dong B, O'Neill Z, Luo D, Bailey (2013) Development and calibration of a reduced-order energy performance model for a mixed-use building. https://doi.org/10.26868/25222708.2013.1412
7. Homod RZ (2013) Review on the HVAC system modeling types and the shortcomings of their application. J Energy 2013:1–10. https://doi.org/10.1155/2013/768632
8. Wu S, Sun J-Q (2012) A physics-based linear parametric model of room temperature in office buildings. Build Environ 50:1–9. https://doi.org/10.1016/j.buildenv.2011.10.005

9. Mossolly M, Ghali K, Ghaddar N (2009) Optimal control strategy for a multi-zone air conditioning system using a genetic algorithm. Energy 34:58–66. https://doi.org/10.1016/j.energy.2008.10.001
10. Tashtoush B, Molhim M, Al-Rousan M (2005) Dynamic model of an HVAC system for control analysis. Energy 30:1729–1745. https://doi.org/10.1016/j.energy.2004.10.004
11. Pedersen CO (2009) A fundamental approach to cooling load calculations for UFADs and stratified spaces. International Building Performance Simulation Association, Glasgow, Scotland, pp 704–708
12. Zajic I, Larkowski T, Sumislawska M, Burnham KJ, Hill D (2011) Modelling of an air handling unit: a Hammerstein-bilinear model identification approach. In: 2011 21st international conference on systems engineering. IEEE, pp 59–63. https://doi.org/10.1109/ICSEng.2011.19
13. Beghi A, Cecchinato L, Paggiaro F, Rampazzo M (2011) VAVAC systems modeling and simulation for FDD applications. In: 2011 9th IEEE international conference on control and automation (ICCA). IEEE, pp 800–805. https://doi.org/10.1109/ICCA.2011.6138039
14. Yan X, Ren Q, Meng Q (2010) Global optimization of VAV air conditioning system. In: 2010 8th world congress on intelligent control and automation. IEEE, pp 5077–5081. https://doi.org/10.1109/WCICA.2010.5554517
15. Cui P, Yang H, Spitler JD, Fang Z (2008) Simulation of hybrid ground-coupled heat pump with domestic hot water heating systems using HVACSIM+. Energy Build 40:1731–1736. https://doi.org/10.1016/j.enbuild.2008.03.001
16. Wemhoff AP, Frank MV (2010) Predictions of energy savings in HVAC systems by lumped models. Energy Build 42:1807–1814. https://doi.org/10.1016/j.enbuild.2010.05.017
17. Park M-H, Shin E-G, Lee H-R, Suh I-S (2010) Dynamic model and control algorithm of HVAC system for OLEV ® application. In: ICCAS 2010. IEEE, pp 1312–1317. https://doi.org/10.1109/ICCAS.2010.5670291
18. Vakiloroaya V, Samali B, Madadnia J, Ha QP (2011) Component-wise optimization for a commercial central cooling plant. In: IECON 2011—37th annual conference of the IEEE Industrial Electronics Society. IEEE, pp 2769–2774. https://doi.org/10.1109/IECON.2011.6119750
19. Jin G-Y, Cai W-J, Lu L, Lee EL, Chiang A (2007) A simplified modeling of mechanical cooling tower for control and optimization of HVAC systems. Energy Convers Manag 48:355–365. https://doi.org/10.1016/j.enconman.2006.07.010
20. Huang G, Wang S, Xu X (2010) Robust Model predictive control of VAV air-handling units concerning uncertainties and constraints. HVAC&R Res 16:15–33. https://doi.org/10.1080/10789669.2010.10390890
21. Huang G (2011) Model predictive control of VAV zone thermal systems concerning bilinearity and gain nonlinearity. Control Eng Pract 19:700–710. https://doi.org/10.1016/j.conengprac.2011.03.005
22. Rehrl J, Horn M (2011) Temperature control for HVAC systems based on exact linearization and model predictive control. In: 2011 IEEE international conference on control applications (CCA). IEEE, pp 1119–1124. https://doi.org/10.1109/CCA.2011.6044437
23. Bi Q, Cai W-J, Wang Q-G, Hang C-C, Lee E-L, Sun Y, Liu K-D, Zhang Y, Zou B (2000) Advanced controller auto-tuning and its application in HVAC systems. Control Eng Pract 8:633–644. https://doi.org/10.1016/S0967-0661(99)00198-7
24. Olivier M, Dominique M (1999) Fault detection and diagnosis on HVAC variable air volume system using artificial neural networks. In: Proceedings of IBPSA building simulation
25. Ding MH, Wang BL, Li L, Zheng YF (2009) Structure and properties of TaCx coating on biomedical 316L stainless steel by RF magnetron sputtering. Key Eng Mater 419–420:537–540. https://doi.org/10.4028/www.scientific.net/KEM.419-420.537
26. Xuemei L, Lixing D, Yuyuan D, Lanlan L (2010) Hybrid support vector machine and ARIMA model in building cooling prediction. In: 2010 international symposium on computer, communication, control and automation (3CA). IEEE, pp 533–536. https://doi.org/10.1109/3CA.2010.5533864

27. Xuemei L, Lixing D, Lanlan L (2010) A novel building cooling load prediction based on SVR and SAPSO. In: 2010 international symposium on computer, communication, control and automation (3CA). IEEE, pp 528–532. https://doi.org/10.1109/3CA.2010.5533863
28. Wang Y, You S, Zhang H, Zheng W, Zheng X, Miao Q (2017) Hydraulic performance optimization of meshed district heating network with multiple heat sources. Energy 126:603–621. https://doi.org/10.1016/j.energy.2017.03.044
29. Zhang S, Gu W, Qiu H, Yao S, Pan G, Chen X (2021) State estimation models of district heating networks for integrated energy system considering incomplete measurements. Appl Energy 282:116105. https://doi.org/10.1016/j.apenergy.2020.116105
30. Wang J, Zhao T (2022) Medium dynamic diffusion characteristics in the pipeline network of variable flow HVAC system towards online decoupling control oriented. Energy Build 258:111827. https://doi.org/10.1016/j.enbuild.2021.111827
31. Lu L, Cai W, Soh YC, Xie L, Li S (2004) HVAC system optimization—condenser water loop. Energy Convers Manag 45:613–630. https://doi.org/10.1016/S0196-8904(03)00181-X
32. Liu Z, Song F, Jiang Z, Chen X, Guan X (2014) Optimization based integrated control of building HVAC system. Build Simul 7:375–387. https://doi.org/10.1007/s12273-014-0161-z
33. Zhuang L, Chen X, Guan X (2020) A decentralized method for energy conservation of an HVAC system. Build Simul 13:155–170. https://doi.org/10.1007/s12273-019-0569-6
34. Gao G, Li J, Wen Y (2020) DeepComfort: energy-efficient thermal comfort control in buildings via reinforcement learning. IEEE Internet Things J 7:8472–8484. https://doi.org/10.1109/JIOT.2020.2992117
35. Jahedi G, Ardehali MM (2012) Wavelet based artificial neural network applied for energy efficiency enhancement of decoupled HVAC system. Energy Convers Manag 54:47–56. https://doi.org/10.1016/j.enconman.2011.10.005
36. Sun R, Liu C, Liu R (2015) Procedure for calculating building chilled and hot water differential pressure setpoint. In: Advances in intelligent systems research, Shenyang, pp 551–554
37. Song SW (2016) An enhanced differential pressure reset control for a central heating system with a variable speed pump. In: 2nd annual international conference on advanced materials, mechanical and structural engineering (AMMSE), Jeju, South Korea, pp 215–218
38. Attaran SM, Yusof R, Selamat H (2015) Enhancement of control's parameter of decoupled HVAC system via adaptive controller through the system identification tool box. J Teknol 76:261–272. https://doi.org/10.11113/jt.v76.4108
39. Shen Y, Cai WJ, Li S (2010) Normalized decoupling control for high-dimensional MIMO processes for application in room temperature control HVAC systems. Control Eng Pract 18:652–664. https://doi.org/10.1016/j.conengprac.2010.03.006
40. Huaguang Z, Cai L (2002) Decentralized nonlinear adaptive control of an HVAC system. IEEE Trans Syst Man Cybern Part C Appl Rev 32:493–498. https://doi.org/10.1109/TSMCC.2002.807271
41. Marmaras J, Burbank J, Kosanovic DB (2016) Primary-secondary de-coupled ground source heat pump systems coefficient of performance optimization through entering water temperature control. Appl Therm Eng 96:107–116. https://doi.org/10.1016/j.applthermaleng.2015.10.027
42. Pardo N, Montero Á, Martos J, Urchueguía JF (2010) Optimization of hybrid—ground coupled and air source—heat pump systems in combination with thermal storage. Appl Therm Eng 30:1073–1077. https://doi.org/10.1016/j.applthermaleng.2010.01.015
43. Zhao T, Wang J, Xu M, Li K (2019) An online predictive control method with the temperature based multivariable linear regression model for a typical chiller plant system. Build Simul. https://doi.org/10.1007/s12273-019-0576-7
44. Camponogara E, Scherer H, Biegler L, Grossmann I (2021) Hierarchical decompositions for MPC of resource constrained control systems: applications to building energy management. Optim Eng 22:187–215. https://doi.org/10.1007/s11081-020-09506-x
45. Lara BGV, Molina LMC, Yanes JPM (2015) Modeling and identification of the cooling dynamics of a tropical island hotel. Energy Build 92:19–28. https://doi.org/10.1016/j.enbuild.2015.01.036

46. Yuan Y, Zhu N, Zhou H, Wang H (2021) A new model predictive control method for eliminating hydraulic oscillation and dynamic hydraulic imbalance in a complex chilled water system. Energies (Basel) 14:3608. https://doi.org/10.3390/en14123608
47. Afram A, Janabi-Sharifi F (2014) Review of modeling methods for HVAC systems. Appl Therm Eng 67:507–519. https://doi.org/10.1016/j.applthermaleng.2014.03.055
48. Brandi S, Piscitelli MS, Martellacci M, Capozzoli A (2020) Deep reinforcement learning to optimise indoor temperature control and heating energy consumption in buildings. Energy Build 224:110225. https://doi.org/10.1016/j.enbuild.2020.110225
49. Zhu J, Yang Q, Xu X, Lu J (2015) A LPV model-based chilled water temperature controller for HVAC systems. Build Serv Eng Res Technol 36:368–385. https://doi.org/10.1177/0143624414555811
50. Fong KF, Hanby VI, Chow TT (2006) HVAC system optimization for energy management by evolutionary programming. Energy Build 38:220–231. https://doi.org/10.1016/j.enbuild.2005.05.008
51. Su C-L, Yu K-T (2013) Evaluation of differential pressure setpoint of chilled water pumps in clean Room HVAC systems for energy savings in high-tech industries. IEEE Trans Ind Appl 49:1015–1022. https://doi.org/10.1109/TIA.2013.2251992
52. Tesiero RC, Nassif N, Gokaraju B, Doss DA (2017) Intelligent approaches for modeling and optimizing HVAC systems' energy use. In: Proceedings of the ASME 11th international conference on energy sustainability, Charlotte
53. Fernandez N, Katipamula S, Wang W, Huang Y, Liu G (2015) Energy savings modelling of re-tuning energy conservation measures in large office buildings. J Build Perform Simul 8:391–407. https://doi.org/10.1080/19401493.2014.961032
54. Trčka M, Hensen JLM (2010) Overview of HVAC system simulation. Autom Constr 19:93–99. https://doi.org/10.1016/j.autcon.2009.11.019
55. Wang S, Ma Z (2008) Supervisory and optimal control of building HVAC systems: a review. HVAC&R Res 14:3–32. https://doi.org/10.1080/10789669.2008.10390991
56. Ma Z, Wang S, Xu X, Xiao F (2008) A supervisory control strategy for building cooling water systems for practical and real time applications. Energy Convers Manag 49:2324–2336. https://doi.org/10.1016/j.enconman.2008.01.019
57. Wang J, Zhang Q, Yoon S, Yu Y (2019) Impact of uncertainties on the supervisory control performance of a hybrid cooling system in data center. Build Environ 148:361–371. https://doi.org/10.1016/j.buildenv.2018.11.026
58. ASHRAE (2019) Supervisory control strategies and optimization. In: ASHRAE handbook—HVAC applications, pp 43.1–43.48
59. Zhao J, Lam KP, Ydstie BE, Loftness V (2016) Occupant-oriented mixed-mode EnergyPlus predictive control simulation. Energy Build 117:362–371. https://doi.org/10.1016/j.enbuild.2015.09.027
60. Cetin KS, Fathollahzadeh MH, Kunwar N, Do H, Tabares-Velasco PC (2019) Development and validation of an HVAC on/off controller in EnergyPlus for energy simulation of residential and small commercial buildings. Energy Build 183:467–483. https://doi.org/10.1016/j.enbuild.2018.11.005
61. Zhang K, Blum D, Cheng H, Paliaga G, Wetter M, Granderson J (2022) Estimating ASHRAE Guideline 36 energy savings for multi-zone variable air volume systems using Spawn of EnergyPlus. J Build Perform Simul 15:215–236. https://doi.org/10.1080/19401493.2021.2021286
62. Qiu A, Yan Z, Deng Q, Liu J, Shang L, Wu J (2020) Modeling of HVAC systems for fault diagnosis. IEEE Access 8:146248–146262. https://doi.org/10.1109/ACCESS.2020.3015526
63. Alibabaei N, Fung AS, Raahemifar K (2016) Development of Matlab-TRNSYS co-simulator for applying predictive strategy planning models on residential house HVAC system. Energy Build 128:81–98. https://doi.org/10.1016/j.enbuild.2016.05.084
64. Cong Y, Hou L, Wu Y, Ma Y (2022) Development of a coupled EnergyPlus-MATLAB simulation based on LSTM for predictive control of HVAC system. Math Probl Eng 2022:1–10. https://doi.org/10.1155/2022/5912967

65. Taheri S, Razban A (2021) Learning-based CO_2 concentration prediction: application to indoor air quality control using demand-controlled ventilation. Build Environ 205:108164. https://doi.org/10.1016/j.buildenv.2021.108164
66. Liu J, Cai W-J, Zhang G-Q (2009) Design and application of handheld auto-tuning PID instrument used in HVAC. In: 2009 4th IEEE conference on industrial electronics and applications. IEEE, pp 1695–1698. https://doi.org/10.1109/ICIEA.2009.5138484
67. Sundriyal N, Yadav P, Chaturvedi M (2017) A novel approach for designing PID controller for set-point tracking for a HVAC process, pp 161–166. https://doi.org/10.1007/978-981-10-1708-7_17
68. Saloux E, Candanedo JA (2020) Optimal rule-based control for the management of thermal energy storage in a Canadian solar district heating system. Sol Energy 207:1191–1201. https://doi.org/10.1016/j.solener.2020.07.046
69. Alimohammadisagvand B, Jokisalo J, Sirén K (2018) Comparison of four rule-based demand response control algorithms in an electrically and heat pump-heated residential building. Appl Energy 209:167–179. https://doi.org/10.1016/j.apenergy.2017.10.088
70. Zhao J, Shan Y (2020) A fuzzy control strategy using the load forecast for air conditioning system. Energies (Basel) 13:530. https://doi.org/10.3390/en13030530
71. Al-Ali AR, Tubaiz NA, Al-Radaideh A, Al-Dmour JA, Murugan L (2012) Smart grid controller for optimizing HVAC energy consumption. In: 2012 international conference on computer systems and industrial informatics. IEEE, pp 1–4. https://doi.org/10.1109/ICCSII.2012.6454548
72. Villar JR, de la Cal E, Sedano J (2009) A fuzzy logic based efficient energy saving approach for domestic heating systems. Integr Comput Aided Eng 16:151–163. https://doi.org/10.3233/ICA-2009-0302
73. Zhao T, Ma L, Zhang J (2015) Experimental study on a novel optimal differential pressure reset method for online application in chilled water system. Procedia Eng 121:714–719. https://doi.org/10.1016/j.proeng.2015.09.013
74. Dubov I (2003) Chilled water plant efficiency. ASHRAE J 45:37–40
75. Chan KT, Yu FW (2004) How chillers react to building loads. ASHRAE J 46:52–63. https://doi.org/10.1016/j.appithermaleng.2003.11.019
76. Chang Y-C (2004) A novel energy conservation method—optimal chiller loading. Electr Power Syst Res 69:221–226. https://doi.org/10.1016/j.epsr.2003.10.012
77. Chang Y-C, Lin J-K, Chuang M-H (2005) Optimal chiller loading by genetic algorithm for reducing energy consumption. Energy Build 37:147–155. https://doi.org/10.1016/j.enbuild.2004.06.002
78. Chang Y-C (2006) An outstanding method for saving energy—optimal chiller operation. IEEE Trans Energy Convers 21:527–532. https://doi.org/10.1109/TEC.2006.871358
79. Zhao T, Zhang J, Ma L (2012) On-line optimization control method based on extreme value analysis for parallel variable-frequency hydraulic pumps in central air-conditioning systems. Build Environ 47:330–338. https://doi.org/10.1016/j.buildenv.2011.07.007
80. Wu Q, Wang X, Shen Q (2016) Research on dynamic modeling and simulation of axial-flow pumping system based on RBF neural network. Neurocomputing 186:200–206. https://doi.org/10.1016/j.neucom.2015.12.064
81. Olszewski P (2016) Genetic optimization and experimental verification of complex parallel pumping station with centrifugal pumps. Appl Energy 178:527–539. https://doi.org/10.1016/j.apenergy.2016.06.084
82. Wegley C, Eusuff M, Lansey K (2000) Determining pump operations using particle swarm optimization. In: Building partnerships. American Society of Civil Engineers, Reston, VA, pp 1–6. https://doi.org/10.1061/40517(2000)206
83. Liu J, Yan F, Yu Q (2009) Research on the modes and strategies of energy-saving for cooling tower fan. Build Sci 25:35–38
84. Taheri S, Hosseini P, Razban A (2022) Model predictive control of heating, ventilation, and air conditioning (HVAC) systems: a state-of-the-art review. J Build Eng 60:105067. https://doi.org/10.1016/j.jobe.2022.105067

85. Hagras H, Packharn I, Vanderstockt Y, McNulty N, Vadher A, Doctor F (2008) An intelligent agent based approach for energy management in commercial buildings. In: 2008 IEEE international conference on fuzzy systems (IEEE world congress on computational intelligence). IEEE, pp 156–162. https://doi.org/10.1109/FUZZY.2008.4630359
86. Beghi A, Cecchinato L, Rampazzo M, Simmini F (2010) Load forecasting for the efficient energy management of HVAC systems. In: 2010 IEEE international conference on sustainable energy technologies (ICSET). IEEE, pp 1–6. https://doi.org/10.1109/ICSET.2010.5684414
87. Zhao L, Huang Z, Fu Q, Fang N, Xing B, Chen J (2023) HVAC optimal control based on the sensitivity analysis: an improved SA combination method based on a neural network. Comput Model Eng Sci 136:2741–2758. https://doi.org/10.32604/cmes.2023.025500
88. Vieira S, Lopez Pinaya WH, Garcia-Dias R, Mechelli A (2020) Deep neural networks. In: Machine learning. Elsevier, pp 157–172. https://doi.org/10.1016/B978-0-12-815739-8.00009-2
89. Dinh HT, Kim D (2022) MILP-based imitation learning for HVAC control. IEEE Internet Things J 9:6107–6120. https://doi.org/10.1109/JIOT.2021.3111454
90. Serale G, Fiorentini M, Capozzoli A, Bernardini D, Bemporad A (2018) Model predictive control (MPC) for enhancing building and HVAC system energy efficiency: problem formulation, applications and opportunities. Energies (Basel) 11:631. https://doi.org/10.3390/en11030631
91. Ham CCW, Singh SPN, Kearney M (2013) Learning-based model predictive control and user feedback in home automation. In: 2013 IEEE/RSJ international conference on intelligent robots and systems. IEEE, pp 2718–2724. https://doi.org/10.1109/IROS.2013.6696740
92. Du Y, Li F, Munk J, Kurte K, Kotevska O, Amasyali K, Zandi H (2021) Multi-task deep reinforcement learning for intelligent multi-zone residential HVAC control. Electr Power Syst Res 192:106959. https://doi.org/10.1016/j.epsr.2020.106959
93. Deng X, Zhang Y, Zhang Y, Qi H (2022) Towards optimal HVAC control in non-stationary building environments combining active change detection and deep reinforcement learning. Build Environ 211:108680. https://doi.org/10.1016/j.buildenv.2021.108680
94. Ala'raj M, Radi M, Abbod MF, Majdalawieh M, Parodi M (2022) Data-driven based HVAC optimisation approaches: a systematic literature review. J Build Eng 46:103678. https://doi.org/10.1016/j.jobe.2021.103678
95. Du Y, Zandi H, Kotevska O, Kurte K, Munk J, Amasyali K, Mckee E, Li F (2021) Intelligent multi-zone residential HVAC control strategy based on deep reinforcement learning. Appl Energy 281:116117. https://doi.org/10.1016/j.apenergy.2020.116117
96. Yang Y, Hu G, Spanos CJ (2022) Stochastic optimal control of HVAC system for energy-efficient buildings. IEEE Trans Control Syst Technol 30:376–383. https://doi.org/10.1109/TCST.2021.3057630
97. Wang Z, Zhao Y, Zhang C, Ma P, Liu X (2022) A general multi agent-based distributed framework for optimal control of building HVAC systems. J Build Eng 52:104498. https://doi.org/10.1016/j.jobe.2022.104498
98. Alhaider M, Fan L, Miao Z (2016) Benders decomposition for stochastic programming-based PV/Battery/HVAC planning. In: 2016 IEEE power and energy society general meeting (PESGM). IEEE, pp 1–5. https://doi.org/10.1109/PESGM.2016.7741775
99. Alhaider M, Fan L (2015) Mixed integer programming for HVACs operation. In: 2015 IEEE power & energy society general meeting. IEEE, pp 1–5. https://doi.org/10.1109/PESGM.2015.7285714
100. Bhattacharya S, Kar K, Chow JH (2019) Economic operation of thermostatic loads under time varying prices: an optimal control approach. IEEE Trans Sustain Energy 10:1960–1970. https://doi.org/10.1109/TSTE.2018.2876797
101. Sankur M, Arnold D, Auslander D (2015) Dynamic programming for optimal load-shedding of office scale battery storage and plug-loads. In: 2015 IEEE power & energy society general meeting. IEEE, pp 1–5. https://doi.org/10.1109/PESGM.2015.7285755
102. Takabatake T, Yamamoto M, Hino H (2022) Algorithm for searching optimal set values of absorption chiller system using Bayesian optimization. Sci Technol Built Environ 28:188–199. https://doi.org/10.1080/23744731.2021.2005376

103. Adhikari R, Pipattanasomporn M, Rahman S (2020) Heuristic algorithms for aggregated HVAC control via smart thermostats for regulation service. IEEE Trans Smart Grid 11:2023–2032. https://doi.org/10.1109/TSG.2019.2946170
104. Garces-Jimenez A, Gomez-Pulido J-M, Gallego-Salvador N, Garcia-Tejedor A-J (2021) Genetic and swarm algorithms for optimizing the control of building HVAC systems using real data: a comparative study. Mathematics 9:2181. https://doi.org/10.3390/math9182181
105. Jun Z, Kanyu Z (2011) A particle swarm optimization approach for optimal design of PID controller for temperature control in HVAC. In: 2011 third international conference on measuring technology and mechatronics automation. IEEE, pp 230–233. https://doi.org/10.1109/ICMTMA.2011.63
106. Sun F, Yu J, Zhao A, Zhou M (2021) Optimizing multi-chiller dispatch in HVAC system using equilibrium optimization algorithm. Energy Rep 7:5997–6013. https://doi.org/10.1016/j.egyr.2021.09.012
107. Caldas LG, Norford LK (2003) Genetic algorithms for optimization of building envelopes and the design and control of HVAC systems. J Sol Energy Eng 125:343–351. https://doi.org/10.1115/1.1591803
108. Yan K, Diduch D, Kaye ME (2019) An improved temperature prediction technique for HVAC units using intelligent algorithms. In: 2019 IEEE energy conversion congress and exposition (ECCE). IEEE, pp 490–494. https://doi.org/10.1109/ECCE.2019.8912944
109. Seong N-C, Kim J-H, Choi W (2019) Optimal control strategy for variable air volume air-conditioning systems using genetic algorithms. Sustainability 11:5122. https://doi.org/10.3390/su11185122
110. Arabali A, Ghofrani M, Etezadi-Amoli M, Fadali MS, Baghzouz Y (2013) Genetic-algorithm-based optimization approach for energy management. IEEE Trans Power Delivery 28:162–170. https://doi.org/10.1109/TPWRD.2012.2219598
111. Ascione F, Bianco N, Mauro GM, Napolitano DF, Vanoli GP (2019) Weather-data-based control of space heating operation via multi-objective optimization: application to Italian residential buildings. Appl Therm Eng 163:114384. https://doi.org/10.1016/j.applthermaleng.2019.114384
112. Papadopoulos S, Kontokosta CE, Vlachokostas A, Azar E (2019) Rethinking HVAC temperature setpoints in commercial buildings: the potential for zero-cost energy savings and comfort improvement in different climates. Build Environ 155:350–359. https://doi.org/10.1016/j.buildenv.2019.03.062
113. Pezeshki Z, Soleimani A, Darabi A (2020) GA-FBHCL: a method for the best HVAC location. In: 2020 zooming innovation in consumer technologies conference (ZINC). IEEE, pp 93–98. https://doi.org/10.1109/ZINC50678.2020.9161819
114. Zhang Y, Zeng P, Zang C (2014) Multi-objective optimal control algorithm for HVAC based on particle swarm optimization. In: Fifth international conference on intelligent control and information processing. IEEE, pp 417–423. https://doi.org/10.1109/ICICIP.2014.7010290
115. Yang R, Wang L (2015) Control strategy optimization for energy efficiency and comfort management in HVAC systems. In: 2015 IEEE power & energy society innovative smart grid technologies conference (ISGT). IEEE, pp 1–5. https://doi.org/10.1109/ISGT.2015.7131863
116. Kusiak A, Xu G, Tang F (2011) Optimization of an HVAC system with a strength multi-objective particle-swarm algorithm. Energy 36:5935–5943. https://doi.org/10.1016/j.energy.2011.08.024
117. Hupp W, Hasandka A, de Carvalho RS, Saleem D (2020) Module-OT: a hardware security module for operational technology. In: 2020 IEEE Texas power and energy conference (TPEC). IEEE, pp 1–6. https://doi.org/10.1109/TPEC48276.2020.9042540
118. Mu B, Li Y, House JM, Salsbury TI (2017) Real-time optimization of a chilled water plant with parallel chillers based on extremum seeking control. Appl Energy 208:766–781. https://doi.org/10.1016/j.apenergy.2017.09.072
119. Karami M, Wang L (2018) Particle Swarm optimization for control operation of an all-variable speed water-cooled chiller plant. Appl Therm Eng 130:962–978. https://doi.org/10.1016/j.applthermaleng.2017.11.037

120. Chen Z, Xu P, Feng F, Qiao Y, Luo W (2021) Data mining algorithm and framework for identifying HVAC control strategies in large commercial buildings. Build Simul 14:63–74. https://doi.org/10.1007/s12273-019-0599-0
121. Wang S, Xing J, Jiang Z, Dai Y (2019) A decentralized swarm intelligence algorithm for global optimization of HVAC system. IEEE Access 7:64744–64757. https://doi.org/10.1109/ACCESS.2019.2913359
122. Liu X, Ren M, Yang Z, Yan G, Guo Y, Cheng L, Wu C (2022) A multi-step predictive deep reinforcement learning algorithm for HVAC control systems in smart buildings. Energy 259:124857. https://doi.org/10.1016/j.energy.2022.124857
123. Chen CL, Chang YC, Chan TS (2014) Applying smart models for energy saving in optimal chiller loading. Energy Build 68:364–371. https://doi.org/10.1016/j.enbuild.2013.04.030
124. Afroz Z, Shafiullah GM, Urmee T, Shoeb MA, Higgins G (2022) Predictive modelling and optimization of HVAC systems using neural network and particle swarm optimization algorithm. Build Environ 209:108681. https://doi.org/10.1016/j.buildenv.2021.108681
125. Zheng Z, Li J, Duan P (2019) Optimal chiller loading by improved artificial fish swarm algorithm for energy saving. Math Comput Simul 155:227–243. https://doi.org/10.1016/j.matcom.2018.04.013
126. Lo C-C, Tsai S-H, Lin B-S (2016) Economic dispatch of chiller plant by improved ripple bee swarm optimization algorithm for saving energy. Appl Therm Eng 100:1140–1148. https://doi.org/10.1016/j.applthermaleng.2016.02.114
127. Yu J, Liu Q, Zhao A, Qian X, Zhang R (2020) Optimal chiller loading in HVAC system using a novel algorithm based on the distributed framework. J Build Eng 28:101044. https://doi.org/10.1016/j.jobe.2019.101044
128. Mozer MC (1998) The neural network house: an environment that adapts to its inhabitants. In: Proceedings of the American Association for Artificial Intelligence spring symposium on intelligent environments, Menlo Park, CA, pp 110–114
129. Biemann M, Scheller F, Liu X, Huang L (2021) Experimental evaluation of model-free reinforcement learning algorithms for continuous HVAC control. Appl Energy 298:117164. https://doi.org/10.1016/j.apenergy.2021.117164
130. Henze G, Schoenmann J (2003) Evaluation of reinforcement learning control for thermal energy storage systems. HVAC&R Res 9:259–275. https://doi.org/10.1080/10789669.2003.10391069
131. Liu S, Henze GP (2005) Evaluation of reinforcement learning for optimal control of building active and passive thermal storage inventory. In: Solar energy. ASMEDC, pp 301–311. https://doi.org/10.1115/ISEC2005-76085
132. Liu S, Henze GP (2006) Experimental analysis of simulated reinforcement learning control for active and passive building thermal storage inventory. Energy Build 38:142–147. https://doi.org/10.1016/j.enbuild.2005.06.002
133. Huang Z, Chen J, Fu Q, Wu H, Lu Y, Gao Z (2020) HVAC optimal control with the multistep-actor critic algorithm in large action spaces. Math Probl Eng 2020:1–12. https://doi.org/10.1155/2020/1386418

Chapter 3
Component Models and Control Loops in HVAC Water System

Before the optimization and control of HVAC water system, some basic definition and principle must be figured out. This chapter provides a description of system configuration, models, and fundamental control in typical HVAC water systems. It establishes the relationship between component operating parameters and energy consumption to provide theoretical support for solving global optimization problems in subsequent steps. The models primarily focus on parallel component, including descriptions of the number of component operating units, chiller, chilled water pump, air handling unit, cooling water pump, and cooling tower. Meanwhile, some conventional methods of tracking setpoint are also presented.

3.1 System Configurations

According to the flow regulation mode of the terminal, the air-conditioning water system can be basically categorized into two types.

The first type is air-conditioning water system with continuous flow regulation. The terminal in this system is equipped with motorized valve. The typical example of the terminal is air handling unit (AHU).

The other type is air-conditioning water system with on–off flow regulation. The terminal is equipped with motorized and pneumatic two-way valve. The typical example of the terminal is fan coiling unit (FCU).

In the following of this section, the examples of these two types of water system will be given, respectively.

Figure 3.1 shows a typical water system with continuous flow regulation composed of c chillers, m pumps, n AHUs in this system. $Q_{\mathrm{WS}i}$ and $Q_{\mathrm{WR}i}$ are flow rate for water supply trunk pipe and return trunk pipe flow in section i, respectively (m^3/s). $Q_{\mathrm{W}i}$ refers to the chilled water flow rate of branch i (m^3/s). $S_{\mathrm{S}i}$ and $S_{\mathrm{R}i}$ denote

T. Zhao et al., *Thermal Distribution Principle and Nonlinear Optimal Control for HVAC Water System*, Indoor Environment and Sustainable Building,
https://doi.org/10.1007/978-981-96-0377-0_3

the impedance of the trunk pipe for supply and return water in section i, respectively (s^2/m^5). S_{Bi} and S_{Vi} respectively represent the impedance of the pipe section (including the cooling coil) and the valve impedance of branch i, s^2/m^5. δ_{Vi} represents the valve position of the regulating valve in branch i. t_{WOi} is the outlet temperature of the cooling coil in branch i, °C. t_{SAi} and t_{RAi} represent the supply air temperature and return air temperature of AHU-i, °C, $i = 1, 2, \ldots, n$.

Figure 3.2 shows a primary pumping on–off control water system as an example. The pipe network consists of n risers, with m branches on each riser. FCU is the branch user, and FCU_{ji} means that the FCU is in the ith branch of the jth riser of the system ($i \in [1, m], j \in [1, n]$). v_{ji} represents the electric two-way valve corresponding to the fan coil. V_j is the continuous regulating valve on the jth riser. t_{WRj} is the chilled water return temperature of the jth riser (°C).

Two common configurations of the HVAC water systems are shown in Fig. 3.3, where subplot (a) shows that the same type components are connected in parallel,

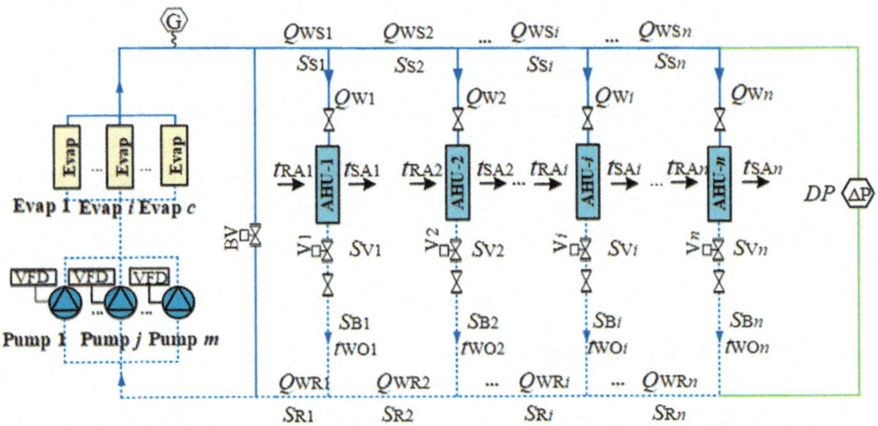

Fig. 3.1 Schematic diagram of typical water system with continuous flow regulation (redrawn from Ref. [1], copyright 2016, with permission from ELSEVIER)

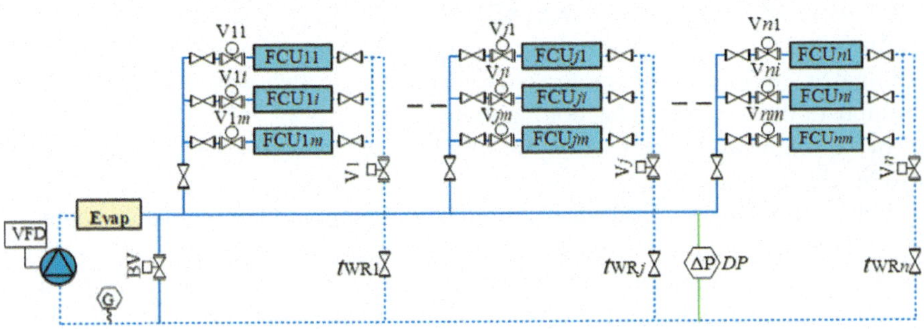

Fig. 3.2 Schematic diagram of typical water system with on–off flow regulation

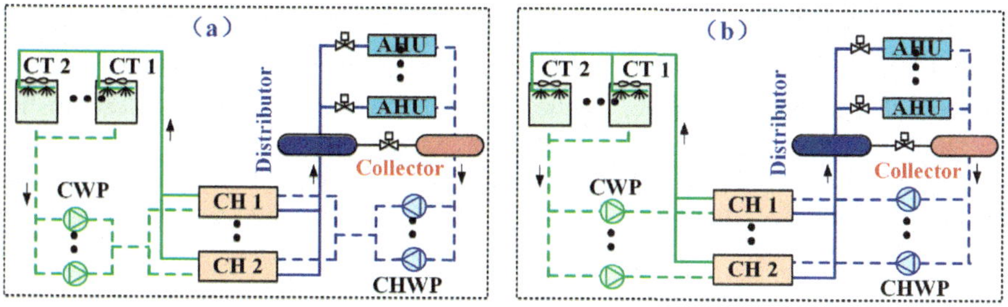

Fig. 3.3 Two common configurations for HVAC water system

while subplot (b) shows that the pump and chiller are connected in series. It can be seen that for the connection form shown in subplot (b), the ON/OFF state of the water pump is synchronized with the corresponding chiller, and each combination can be regarded as a one-chiller-one-pump system.

3.2 Component Models

3.2.1 Description of Component Numbers

There are three different descriptions related to the number of component, namely, the total number of units (denoted as N_{comp}), the number of units in operation (denoted as $N_{\mathrm{r,comp}}$), and the units on–off status (denoted as $N_{\mathrm{d,comp}}$). Among these, N_{comp} and $N_{\mathrm{r,comp}}$ are integer variables, while $N_{\mathrm{d,comp}}$ is a discrete vector composed of 0 and 1, representing the on–off status of each device as shown in Eq. (3.1). In this equation, $N_{\mathrm{d,comp},i}$ is a byte-type data with values of 0 or 1, indicating the on–off status of the ith unit in parallel component of type comp, where 0 represents unit shutdown, and 1 represents unit operating. The subscript comp represents the component type, with values from the set {ch, chwp, cwp, ct, ahu}, representing chiller, chilled water pump, cooling water pump, cooling tower, and air handling unit (AHU), respectively.

$$
\begin{aligned}
N_{\mathrm{d,comp}} = \left(N_{\mathrm{d,comp},1}, \ldots, N_{\mathrm{d,comp},b}, \ldots, N_{\mathrm{d,comp},N_{\mathrm{comp}}}\right)&, \\
\mathrm{comp} \in \{\mathrm{ch, chwp, cwp, ct, ahu}\}& \qquad (3.1)
\end{aligned}
$$

$$
N_{\mathrm{r,comp}} = \sum_{b=1}^{N_{\mathrm{comp}}} N_{\mathrm{d,comp},b}. \qquad (3.2)
$$

Taking the chiller unit as an example, N_{ch} represents the total number of chiller units in the system, $N_{\mathrm{r,ch}}$ represents the number of chiller units currently in operation, and $N_{\mathrm{d,ch},1}$ represents the operational status of the first chiller unit.

3.2.2 Chiller Model

The calculation of chiller power can be referenced using a simple model established by Stoecker [2], expressed in the form of the rated power multiplied by correction factors. By incorporating the information regarding the on–off status of the component, denoted as $N_{d,ch}$, the power model for chiller units can be formulated as shown in Eq. (3.3).

$$P_{ch} = \sum_{i=1}^{N_{ch}} \left(P_{ch,R,i} \cdot PLR_{ch,f,i} \cdot T_{ch,f,i} \cdot N_{d,ch,i} \right). \tag{3.3}$$

In this equation, $PLR_{ch,f}$ is a correction factor for the partial load ratio of the chiller units, related to the cooling capacity (Q_{ch}), and can be calculated using Eq. (3.4). $T_{ch,f}$ represents the temperature correction factor for the chiller units, associated with the evaporator outlet temperature ($T_{E,o}$) and condenser outlet temperature ($T_{C,o}$), which can be obtained through Eq. (3.5). Coefficients a_0–a_2 and b_0–b_5 can be obtained through data fitting, with fitting data available in the equipment manufacturer's documentation or variable operating condition correction factor tables.

$$PLR_{ch,f,i} = a_{0,i} + a_{1,i}\left(\frac{Q_{ch,i}}{Q_{ch,R,i}}\right) + a_{2,i}\left(\frac{Q_{ch,i}}{Q_{ch,R,i}}\right)^2, \tag{3.4}$$

$$T_{ch,f,i} = b_{0,i} + b_{1,i}T_{E,o} + b_{2,i}T_{E,o^2} + b_{3,i}T_{C,o} + b_{4,i}T_{C,o^2} + b_{5,i}T_{E,o}T_{C,o}. \tag{3.5}$$

It is evident that the power of the chiller units (P_{ch}) is primarily related to Q_{ch}, $T_{E,o}$, and $T_{C,o}$.

3.2.3 Chilled Water Pump Model

The power of chilled water pump can be calculated based on the flow rate (m_{chw}) and the head (H_{chw}). Therefore, the power model for parallel chilled water pumps can be represented by Eq. (3.6).

$$P_{chwp} = \sum_{j=1}^{N_{chwp}} \left(\frac{m_{chw,j} \cdot H_{chw,j}}{g_c \cdot \eta_{chw,j}} \cdot N_{d,chwp,j} \right). \tag{3.6}$$

In this equation, g_c is a constant related to the units of m and H, with the gravitational acceleration g taken as 9.8 m/s^2 and power unit in kW. If the unit of m is in m^3/h and unit of H is in mH$_2$O, then g_c is 367.3. If the unit of m is in kg/s and unit of H is in kPa, then g_c is 134.9. η_{chw} is the efficiency of the chilled water pump, which has a certain functional relationship with the flow rate and head and can be obtained through constructing a simple function as shown in Eq. (3.7) and linear fitting based on operational data.

$$\eta_{\text{chw},j} = f_j(m_{\text{chw},j}, H_{\text{chw},j}). \tag{3.7}$$

Therefore, the power of the chilled water pumps (P_{chw}) is primarily related to m_{chw} and H_{chw}.

For the parallel variable frequency pump set, there are some principles which can be used for both chilled water pump and cooling water pump. Define the ratio of current speed of the pump and the rated speed as the speed ratio, denoted as w. When $w = 1$, i.e., when the pump operates at the rated speed, the pump head, efficiency, and power can be calculated by Eqs. (3.8)–(3.10).

$$H_{\text{PR}} = a_1 Q_{\text{PR}}^2 + a_2 Q_{\text{PR}} + a_3. \tag{3.8}$$

In the equation above, H_{PR} is the head of the pumps operating at the rated speed (m). Q_{PR} is the flow rate of the pumps operating at rated speed (m^3/s). a_1, a_2, and a_3 are the constants of the performance, which can be obtained from sample data or experiment.

$$\eta_{\text{PR}} = b_1 Q_{\text{PR}}^2 + b_2 Q_{\text{PR}} + b_3. \tag{3.9}$$

In the equation above, η_{PR} is the efficiency of the pumps operating at the rated speed (%). b_1, b_2, and b_3 are the constants of the performance, which can be obtained from sample data or experiment.

$$P_{\text{PR}} = \frac{10 H_{\text{PR}} Q_{\text{PR}}}{\eta_{\text{PR}} \eta_{\text{M}} \eta_{\text{VFD}}}. \tag{3.10}$$

In the equation above, P_{PR} represents the power of the pumps operating at the rated speed (kW). η_{M} is the motor efficiency (%) and η_{VFD} is the efficiency of the variable frequency drive.

Based on the characteristics of variable frequency pump and the similarity law, there is:

$$\begin{cases} Q_{\text{PR}} = \frac{Q_{\text{P}}}{w} \\ H_{\text{PR}} = \frac{H_{\text{P}}}{w^2} \\ P_{\text{PR}} = \frac{P_{\text{P}}}{w^3} \\ \eta_{\text{PR}} = \eta_{\text{P}} \end{cases}. \tag{3.11}$$

In Eq. (3.11), Q_{P} is the flow rate of the pump operating at the speed ratio of w (m^3/s). H_{P} is the head of the pump operating at the speed ratio of w (m). P_{P} is the power of the pump operating at the speed ratio of w (kW). η_{P} is the efficiency of the pump operating at the speed ratio of w (%).

Substitute Eq. (3.11) into Eqs. (3.8)–(3.10), there is:

$$H_P = a_1 Q_P^2 + a_2 w Q_P + a_3 w^2, \tag{3.12}$$

$$\eta_P = b_1 w^{-2} Q_P^2 + b_2 w^{-1} Q_P + b_3, \tag{3.13}$$

$$P_P = \frac{10 H_P Q_P}{\eta_P \eta_M \eta_{VFD}}. \tag{3.14}$$

3.2.4 *AHU Model*

Similar to the calculation of chilled water pump power, the power of the AHU fan can also be calculated based on the supply air volume (m_{sa}) and fan pressure (H_{sa}). Therefore, the power model of AHU can be expressed as Eq. (3.15).

$$P_{ahu} = \sum_{k=1}^{N_{ahu}} \frac{m_{sa,k} \cdot H_{sa,k}}{g_c \cdot \eta_{ahu,k}}. \tag{3.15}$$

In this equation, g_c is the same constant as in Eq. (3.6). η_{sa} is the efficiency of the supply air fan, which has a certain functional relationship with the air volume and pressure as shown in Eq. (3.16) and can be obtained through fitting using operational data.

$$\eta_{sa,k} = f_k\left(m_{sa,k}, H_{sa,k}\right). \tag{3.16}$$

The supply air volume of the AHU can be represented as the sum of the supply air volume in all rooms within its responsible area, as shown in Eq. (3.17).

$$m_{sa,k} = \sum_{l=1}^{N_{room}} m_{sa,k,l}. \tag{3.17}$$

Therefore, the power of the AHU supply fan (P_{ahu}) is primarily related to m_{sa} and H_{sa}.

3.2.5 *Cooling Water Pump Model*

The cooling water pump is in an open-loop system, and its power characteristics are slightly different from the chilled water pump and are primarily related to the water flow rate of the cooling water pump, which can be formulated by Eq. (3.18).

$$P_{\text{cwp}} = \sum_{m=1}^{N_{\text{cwp}}} P_{\text{cwp,R},m} N_{\text{d,cwp},m}$$
$$\left[c_{0,m} + c_{1,m}\text{PLR}_{\text{cwp},m} + c_{2,m}\left(\text{PLR}_{\text{cwp},m}\right)^2 + c_{3,m}\left(\text{PLR}_{\text{cwp},m}\right)^3\right]. \tag{3.18}$$

In this equation, c_0–c_3 are model fitting coefficients that can be identified through operational data. $\text{PLR}_{\text{cwp},m}$ represents the partial load ratio of the mth cooling water pump, defined as the ratio of actual water flow to rated flow, as shown in Eq. (3.19).

$$\text{PLR}_{\text{cwp},m} = \frac{m_{\text{cw},m}}{m_{\text{cw,R},m}}. \tag{3.19}$$

Therefore, the power of the cooling water pump (P_{cwp}) is mainly related to m_{cwp}.

3.2.6 *Cooling Tower Model*

Referring to the power model of cooling water pumps, the power model of cooling tower fans can be represented as shown in Eq. (3.20). In this equation, d_0–d_3 are model fitting coefficients that can be identified through operational data. $\text{PLR}_{\text{ct},n}$ represents the partial load ratio of the nth cooling tower fan, defined as the ratio of actual air volume to rated air volume, as shown in Eq. (3.21).

$$P_{\text{ct}} = \sum_{n=1}^{N_{\text{ct}}} P_{\text{ct,R},n} N_{\text{d,ct},n}$$
$$\left[d_{0,n} + d_{1,m}\text{PLR}_{\text{ct},n} + d_{2,n}\left(\text{PLR}_{\text{ct},n}\right)^2 + d_{3,n}\left(\text{PLR}_{\text{ct},n}\right)^3\right], \tag{3.20}$$

$$\text{PLR}_{\text{ct},n} = \frac{\text{m}_{\text{a},n}}{\text{m}_{\text{a,R},n}}. \tag{3.21}$$

Therefore, the power of the cooling tower fan (P_{ct}) is primarily related to m_{a}.

3.2.7 *Pipe Network Model*

The pipe network of the HVAC water system can be regarded as a system composed of several pipes interconnected and combined, and pipelines are the most basic elements of the system. Therefore, before the optimal control of the system, the pipe network model is particularly important and must be formulated. This section will model and analyze the pipeline characteristics under steady-state and non-steady-state conditions and obtain the universal hydraulic and thermal response characteristics of a single pipeline. Then, combined with graph theory, the influence of local resistance components such as valves and bends in the pipe network on hydraulic characteristics, and the influence of heat exchange between the fluid in the pipe and the external environment on thermal characteristics will be comprehensively considered. The model of the pipe network under steady and unsteady conditions will be constructed.

For pipelines under steady-state conditions, the fluid flowrate inside the pipeline does not change. For incompressible fluids, the hydraulic characteristics of the fluid can be described using the total flow Bernoulli equation. If we analyze any two Sects. 3.1 and 3.2 in the pipeline, the following equation is satisfied:

$$z_1 + \frac{p_1}{\gamma} + \frac{v_1^2}{2g} = z_2 + \frac{p_2}{\gamma} + \frac{v_2^2}{2g} + h_{w1,2}, \tag{3.22}$$

where z is the potential energy of the unit weight of fluid passing through the section in the total flow (position height, or head); $\gamma = \rho g$ is the specific weight of the fluid; v is the average velocity of the fluid passing through the section; h is the resistance loss of the fluid between the two sections, including the resistance loss along the pipeline and the local resistance loss.

The resistance loss along the pipeline is generated by the viscosity of the fluid itself and the roughness of the pipe wall. It can be described by the Darcy formula:

$$h_f = f \frac{L}{D} \frac{v^2}{2g}, \tag{3.23}$$

where f is the resistance coefficient along the pipeline, which is related to the roughness of the pipe wall, pipe diameter, and Reynolds number; L is the length of the pipe; D is the diameter of the pipe; v is the average flow velocity of the section.

The local resistance loss can be described as:

$$h_m = \zeta \cdot \frac{v^2}{2g}, \tag{3.24}$$

where ζ is the dimensionless local resistance loss coefficient.

The flow rate of the fluid inside the pipeline satisfies the following relationship with the pressure difference between the two ends of the pipeline:

$$\Delta p = s \cdot q^2, \tag{3.25}$$

where Δp is the pressure difference between the two ends of the pipeline; s is the pipeline impedance; q is the volume flow rate of the pipeline.

The pipeline impedance can be described as [3]:

$$s = \frac{8\rho}{\pi^2 D^4} \sum_{i=1}^{n} \left(f_i \frac{L_i}{D_i} + \zeta_i \right). \tag{3.26}$$

Equation (3.25) can be used to describe the hydraulic characteristics of the pipeline under steady-state conditions.

For the hydraulic characteristics of the pipe network, the method of graph theory [3] can be introduced. In the HVAC water system pipeline network, if the intersection points of each supply pipe section are abstracted as nodes, they are denoted as:

$$V = \{v_1, v_2, \ldots, v_M\}. \tag{3.27}$$

The pipe sections between the intersection points are abstracted as edges, denoted as:

$$E = \{e_1, e_2, \ldots, e_P\}. \tag{3.28}$$

The direction of the edge is determined according to the flow direction of the fluid in the pipe section. Then, the water system pipeline network can be abstracted as a directed graph $G(V, E)$ composed of several nodes and edges. If $I^+(v_i)$ is the set of all edges starting from v_i and $I^-(v_i)$ is the set of all edges ending at v_i, then the relationship between the nodes and edges in the pipeline network diagram can be described using the incidence matrix B, defined as:

$$b_{ij} = \begin{cases} 1 & e_j \in I^+(v_i), \\ -1 & e_j \in I^-(v_i), \\ 0 & \text{Otherwise} \end{cases} \tag{3.29}$$

Through proof, it can be shown that the rank of the incidence matrix B is $M - 1$, that is, any $M - 1$ row vectors in the $M \times P$ order matrix B are linearly independent. If the row vector corresponding to a certain node v_i is removed from the matrix B, the basic incidence matrix B_r with v_i as the reference node can be obtained, which is an $(M - 1) \times P$ order full rank matrix. Through the basic incidence matrix is corresponding to a certain reference node, the relationship between the node flow and the edge flow associated with that reference node, namely the flow balance equation, can be established. Let vector Q be the set of flow rates for each edge, and vector q be the set of net flow rates for each node except the reference node, which can be represented as:

$$Q = \begin{bmatrix} Q_1 \ Q_2 \cdots Q_P \end{bmatrix}^{\mathrm{T}}, \tag{3.30}$$

$$q = \begin{bmatrix} q_1 \ q_2 \cdots q_M \end{bmatrix}^{\mathrm{T}}, \tag{3.31}$$

where Q_i ($i = 1, \ldots, P$) is the flow rate corresponding to edge e_i, and q_j ($j = 1, \ldots, M$) is the net flow rate for each node. The net flow rate of the node is defined as the difference between the total flow into the node and the total flow out of the node connected to the node; therefore, it can be represented by the basic incidence matrix B_r as:

$$q = B_r \cdot Q. \tag{3.32}$$

In the HVAC water system pipeline network, the supply and return water pipeline forms a ring-shaped pipeline network. Without considering flow loss, the net flow of each node in the pipeline network diagram can be regarded as 0, and the above equation can be simplified to:

$$B_r \cdot Q = 0. \tag{3.33}$$

The set of closed loops formed by several closed edges in the pipeline network is called the basic loop, and its collection is denoted as:

$$C = \{c_1, c_2, \ldots, c_V\}, \tag{3.34}$$

where L is the number of loops, and the following relationship can be proved in the ring-shaped pipeline network diagram:

$$V = P - (M - 1). \tag{3.35}$$

If $C^+(c_i)$ is the set of all edges that form c_i and have the same direction as c_i, and $C^-(c_i)$ is the set of all edges that form c_i but have the opposite direction as c_i, then the basic loop matrix C_r can be described as:

$$C_r = \left(C_{ij}\right)_{V \times P}, \quad C_{ij} = \begin{cases} 1 & e_j \in C^+(c_i), \\ -1 & e_j \in C^-(c_i), \\ 0 & \text{Otherwise} \end{cases} \tag{3.36}$$

There is the following relationship between B_r and C_r in the ring-shaped pipeline network:

$$B_r C_r^T = 0. \tag{3.37}$$

Based on the above definition, all topological relationships of the directed graph $G(V, E)$ can be described. Figure 3.4 is an example of a simple ring-shaped pipeline network graph model. According to the above definition, its adjacency matrix and basic loop matrix can be represented as:

$$B = \begin{bmatrix} 1 & 0 & -1 & 0 & 0 & 0 & 0 \\ -1 & 1 & 0 & 1 & 0 & 0 & 0 \\ 0 & -1 & 0 & 0 & 1 & 0 & 0 \\ 0 & 0 & 1 & 0 & 0 & -1 & 0 \\ 0 & 0 & 0 & -1 & 0 & 1 & -1 \\ 0 & 0 & 0 & 0 & -1 & 0 & 1 \end{bmatrix}, \tag{3.38}$$

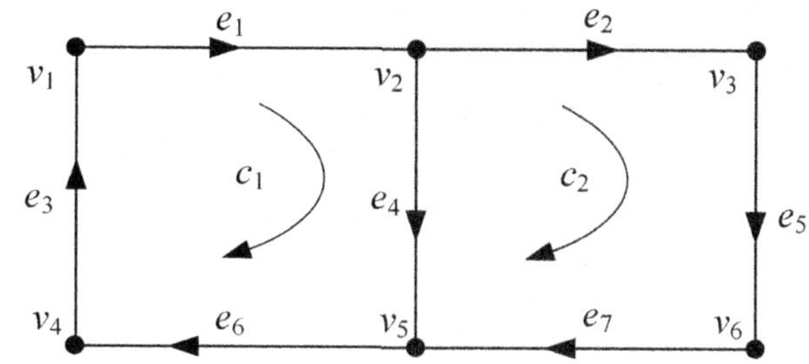

Fig. 3.4 Example for the graph model

$$C_r = \begin{bmatrix} 1 & 0 & 1 & 1 & 0 & 1 & 0 \\ 0 & 1 & 0 & -1 & 1 & 0 & 1 \end{bmatrix}. \tag{3.39}$$

The chilled water pipeline network diagram is a connected graph. If the connected graph is traversed, the set of all nodes and edges passed through in the process can be regarded as a tree, called the spanning tree T. The spanning tree contains all the nodes and connecting edges (branches) in the pipeline network, there is only one path between any two adjacent nodes, and it does not contain any loops. The edges other than the spanning tree in the connected graph are called the surplus branches. It can be proved that for a ring-shaped pipeline network diagram with M nodes and P edges, the number of branches is $(M - 1)$, and the number of surplus branches is $(P - M + 1)$. The set of pipeline pressure drops for all edges in the pipeline network diagram is denoted as Δp.

$$\Delta p = \left[\, \Delta p_1 \ \Delta p_2 \ \cdots \ \Delta p_P \,\right]^{\mathrm{T}}. \tag{3.40}$$

In the formula, Δp_i can be calculated by Eq. (3.25). Since the overall pressure drop within each basic loop is 0, the following pressure balance equation exists:

$$C_r \cdot \left(\Delta p - H_{\mathrm{p}}\right) = 0, \tag{3.41}$$

where H_{p} represents the water pump head vector in the HVAC water system pipe network and can be obtained from the pump flow head characteristic curve. If the branch has no pump, the corresponding value is 0. For the flow balance equation (3.33), there are $(M - 1)$ equations and P variables; for the pressure balance equation (3.41), there are $(P - M + 1)$ equations. By combining Eqs. (3.33) and (3.41), a nonlinear equation system $F(Q)$ with P equations and variables can be obtained, and the hydraulic model can be solved by solving this nonlinear equation system; $F(Q)$ can be denoted as:

$$F(Q) = \begin{cases} B_r \cdot Q \\ C_r \cdot \left(\Delta p - H_p\right) \end{cases}. \tag{3.42}$$

Through the previous analysis, when $F(Q) = 0$, it is the steady-state hydraulic model of the HVAC water system pipeline network. For such a nonlinear equation system, it can be solved by the Newton iteration method [3]. Assuming Q^k is an approximate root of $F(Q)$, the equation can be expanded at Q^k, obtaining the following form:

$$F(Q) = F(Q^k) + F'(Q^k) \cdot (Q - Q^k) + o(Q - Q^k), \tag{3.43}$$

where $F'(Q^k)$ is the differential of F with respect to Q, and the expression is as follows:

$$F'(Q) = \begin{cases} B_r \\ C_r \cdot \left(\frac{\partial \Delta p}{\partial Q} - \frac{\partial H_p}{\partial Q}\right) \end{cases}. \tag{3.44}$$

Ignoring the infinitesimal term $o(Q - Q^k)$, the expansion can be approximated as:

$$F(Q) = F(Q^k) + F'(Q^k) \cdot (Q - Q^k). \tag{3.45}$$

Let $F(Q) = 0$, the following equation can be obtained:

$$F'(Q^k) \cdot (Q - Q^k) = -F(Q^k). \tag{3.46}$$

According to the Newton iteration method, the following iterative relationship can be derived:

$$Q^{k+1} = Q^k - F'(Q^k)^{-1} \cdot F(Q^k), \tag{3.47}$$

where Q^k is the flow vector of all edges at the kth iteration, and in order to facilitate the iterative calculation of the language program, the iteration termination condition is set as:

$$\Delta Q^k = Q^{k+1} - Q^k, \quad \left\| \Delta Q_2^k \right\| < \varepsilon, \tag{3.48}$$

where ε is the residual limit and can be set as 10^{-6}.

3.3 Control Loops

To gain a clear understanding of the fundamental control principles and control loops in the HVAC water systems, the following will provide an introduction to common control loops in the system, serving as a reference for subsequent control optimization. Control loops in HVAC water system primarily consist of the control loops in cooling water system and chilled water system. The following will provide an introduction to the control loops in these two subsystems.

3.3.1 Control Loops in Cooling Water System

The cooling water control system consists of the following two control loops:

(1) Control loop of cooling tower supply and return water temperature difference:

The control schematic and procedure of this control loop are depicted in Figs. 3.5 and 3.6, respectively. The specific description is as follows:

Control variable: Differential temperature of cooling tower supply and return water.
Control device: Cooling water pump with variable frequency drive.
Control loop function: By adjusting the frequency of the water pump variable frequency drive to change the speed of the cooling water pump, it regulates the cooling water flow to ensure the setpoint of the water temperature difference between cooling tower inlet and outlet.

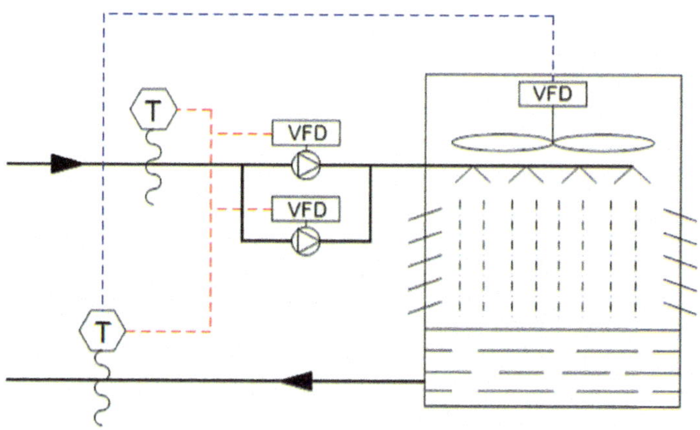

Fig. 3.5 Control schematic of cooling tower differential temperature

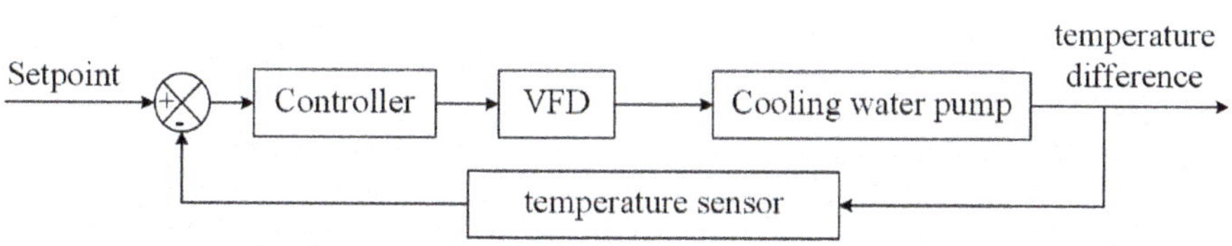

Fig. 3.6 Block diagram of cooling tower differential temperature

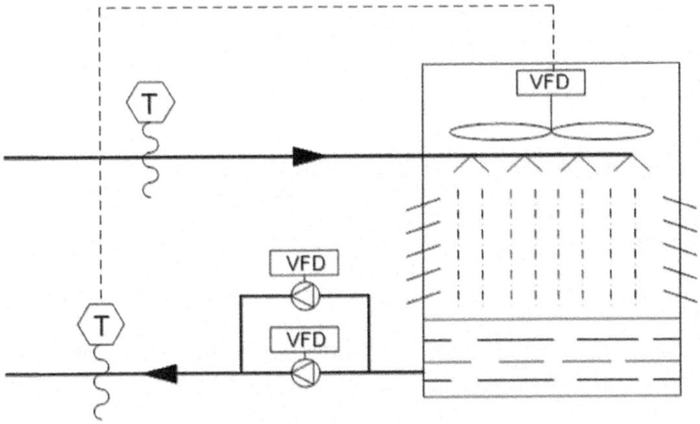

Fig. 3.7 Control schematic of cooling tower outlet temperature

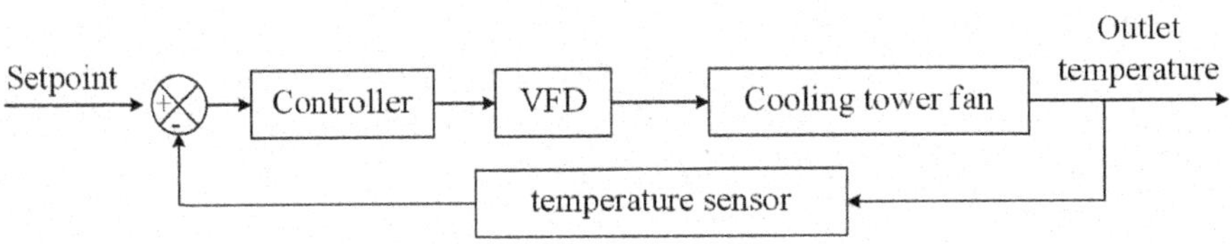

Fig. 3.8 Block diagram of cooling tower outlet temperature

(2) Control loop of cooling tower outlet water temperature:

The control schematic and procedure of this control loop are depicted in Figs. 3.7 and 3.8, respectively. The specific description is as follows:

Control variable: Supply water temperature of cooling tower.
Control device: Cooling tower fan with variable frequency drive.
Control loop function: By adjusting the frequency of the variable frequency drive of the cooling tower fan to change the speed of the fan, it regulates the air volume of cooling air to ensure the setpoint of the cooling tower outlet water temperature.

3.3.2 Control Loops in Chilled Water System

The chilled water control system consists of the following three control loops:

(1) Control loop of pressure differential between chilled water pump inlet and outlet:

The control schematic and procedure of this control loop are depicted in Figs. 3.9 and 3.10, respectively. The specific description is as follows:

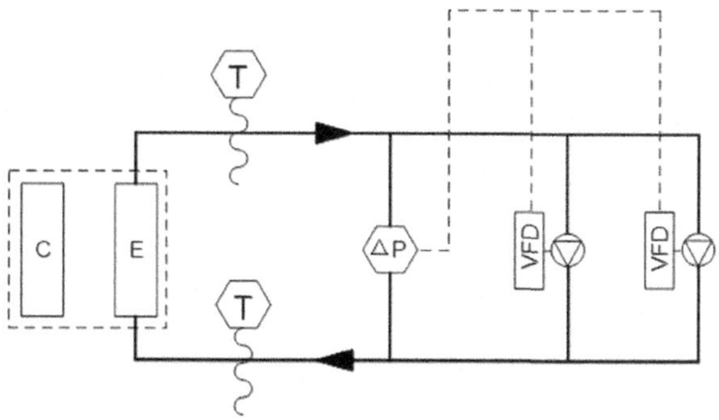

Fig. 3.9 Control schematic of chilled water pump pressure differential

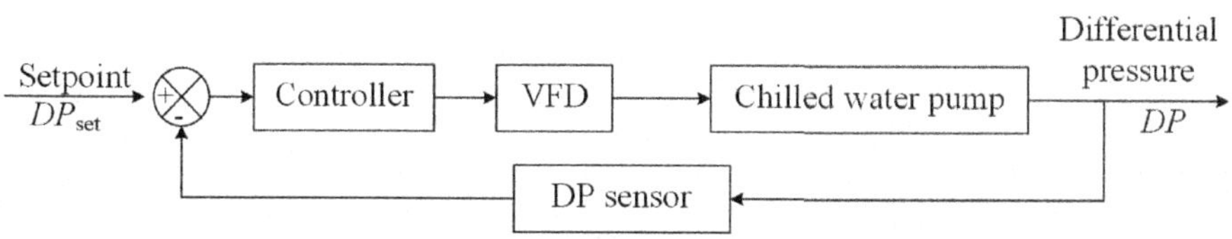

Fig. 3.10 Block diagram of chilled water pump pressure differential

Control variable: Chilled water pump inlet and outlet pressure differential.
Control device: Chilled water pump with variable frequency drive.
Control loop function: By adjusting the frequency of the variable frequency drive to change the speed of the chilled water pump and regulate the chilled water flow rate, and then the setpoint of the chilled water pump inlet and outlet pressure differential will be ensured.

(2) Control loop of temperature difference between chilled water supply and return:

The control schematic and procedure of this control loop are depicted in Figs. 3.11 and 3.12, respectively. The specific description is as follows:

Fig. 3.11 Control schematic of temperature difference control for chilled pump

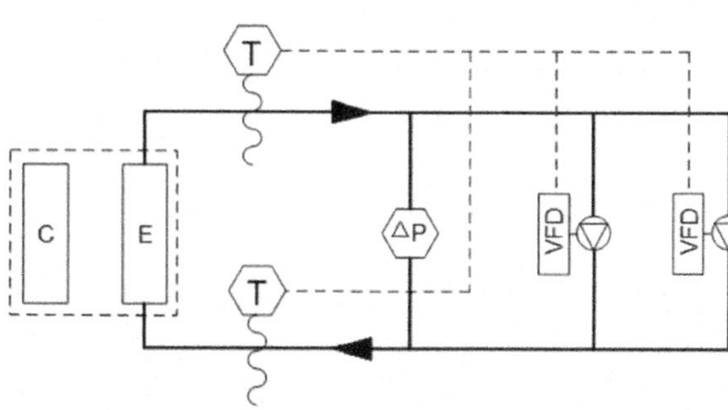

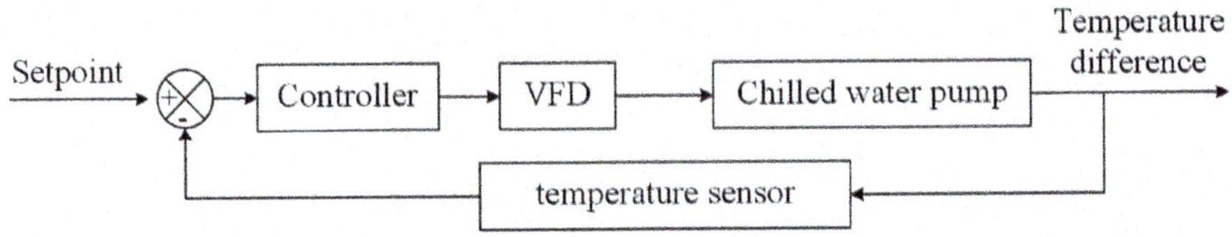

Fig. 3.12 Block diagram of temperature difference control for chilled pump

Control variable: Temperature difference between chilled water supply and return.
Control device: Chilled water pump with variable frequency drive.
Control loop function: By adjusting the frequency of the variable frequency drive to change the speed of the chilled water pump, and regulate the chilled water flow rate, and then ensuring the setpoint of the chilled water supply and return temperature differential.

(3) Chilled water supply temperature control loop

The control schematic and procedure of this control loop are depicted in Figs. 3.13 and 3.14, respectively. The specific description is as follows:

Control variable: Chilled water supply temperature.
Control device: Refrigerant solenoid valve.
Control loop function: By adjusting the opening of the refrigerant solenoid valve to change the refrigerant flow rate through the evaporator, and the evaporator load

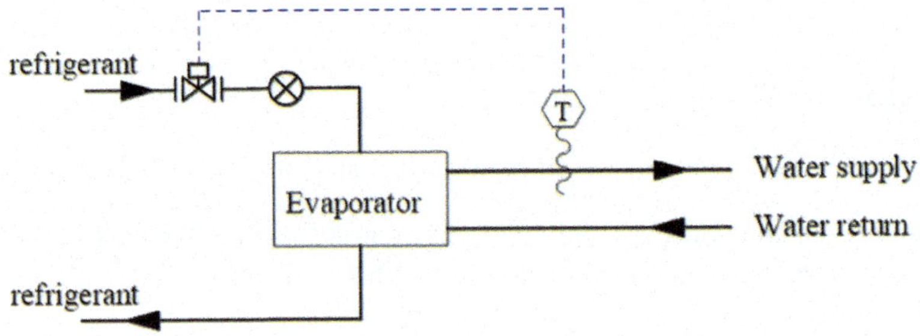

Fig. 3.13 Control schematic of water supplied temperature control

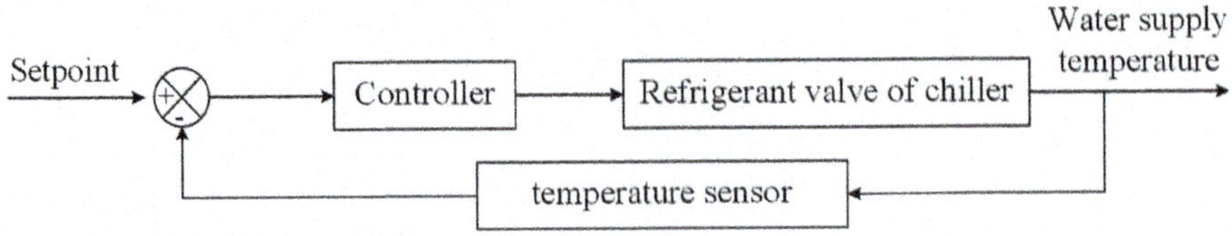

Fig. 3.14 Block diagram of water supplied temperature control

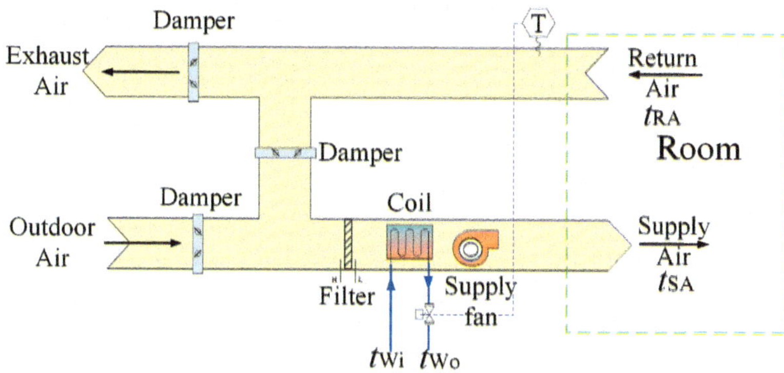

Fig. 3.15 Control schematic of constant air volume AHU return air temperature control

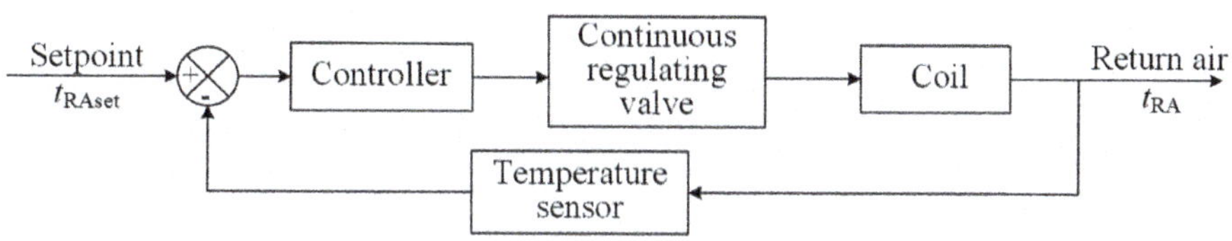

Fig. 3.16 Block diagram of constant air volume AHU return air temperature control loop

will be altered accordingly, and then ensuring the setpoint of the evaporator outlet temperature (chilled water supply temperature).

(4) Constant air volume AHU return air temperature control loop

The control schematic and procedure of this control loop are depicted in Figs. 3.15 and 3.16, respectively. The specific description is as follows:

Control variable: Return air temperature.
Control device: Motorized continuous regulating valve.
Control loop function: By adjusting the opening of the motorized valve to change the water volume passing through the cooling coil based on the differential between the return air temperature and the setpoint, to remain the return air temperature at the setpoint.

(5) Variable air volume AHU return air temperature control loop

The control schematic and procedure of this control loop are depicted in Figs. 3.17 and 3.18, respectively. The specific description is as follows:

Control variable: Return air temperature.
Control device: The fan with variable frequency drive.

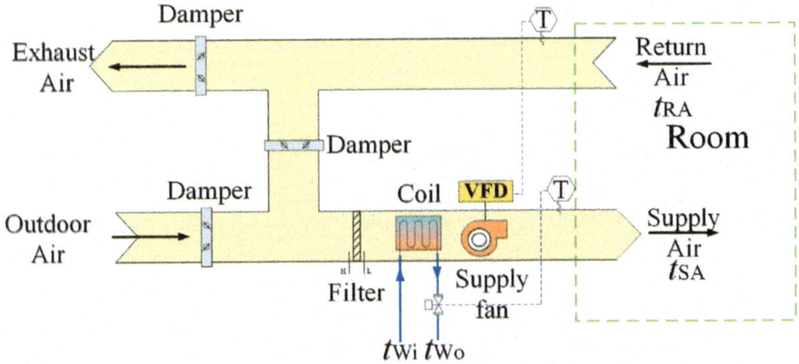

Fig. 3.17 Control schematic of variable air volume AHU return air temperature control

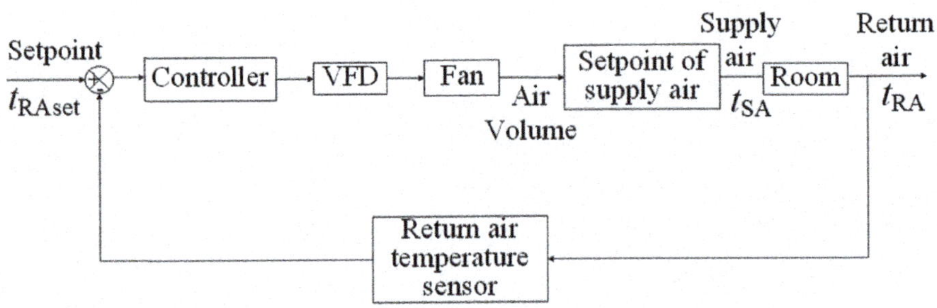

Fig. 3.18 Block diagram of variable air volume AHU return air temperature control loop

Control loop function: By adjusting the fan's speed based on the difference between the return air temperature and the setpoint, which controls the supply air volume entering the room.

Besides, unlike the constant air AHU, the opening of the motorized valve is regulated based on the differential between the sending air temperature and the setpoint.

(6) FCU room temperature control loop

The control schematic and procedure of this control loop can be depicted in Figs. 3.19 and 3.20, respectively. The specific description is as follows:

Control variable: Room temperature.
Control device: On–off valve.
Control loop function: By controlling the opening of the electric on–off valve to maintain the room temperature as the setpoint depending on the deviation of the temperature from its setpoint.

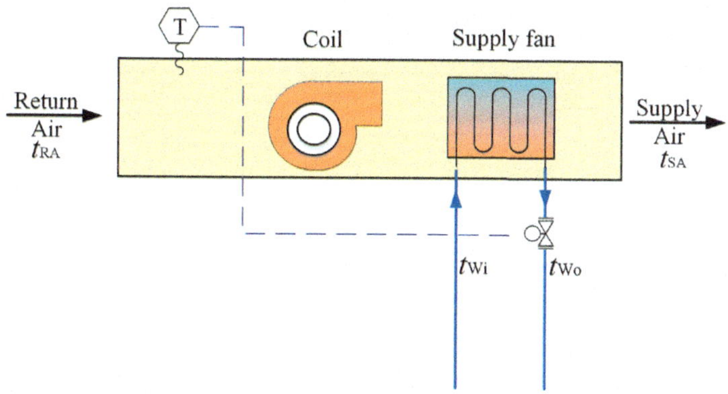

Fig. 3.19 Control schematic of return air temperature control

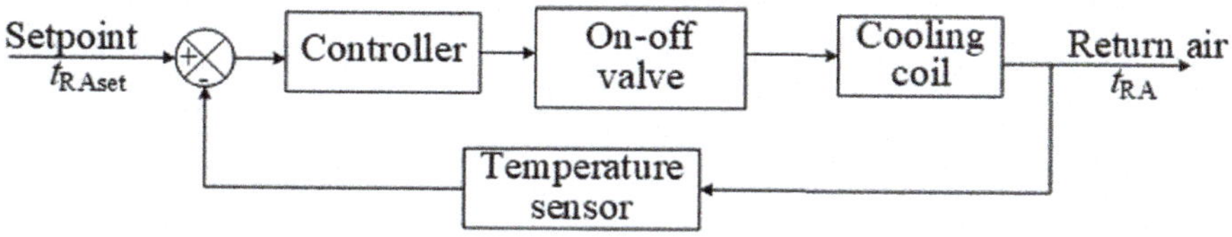

Fig. 3.20 Block diagram of return air temperature control

3.4 Methods of Tracking Control Loop Setpoint

(1) Rule-based control method

Rule-based control strategy is a method of controlling system behavior by guiding it through pre-set rules and conditions. This strategy is usually defined based on expert knowledge and experience and can be applied in various fields such as industrial control, traffic management, and financial risk control. The rule-based control method has the following advantages:

(a) Easy to understand and implement: Rule-based control strategies are usually based on logic and rules, so they are relatively easy to understand and implement.
(b) Higher flexibility: Rules can be flexibly adjusted and modified according to actual situations to adapt to different environments and needs.
(c) Higher reliability: Rules are validated by expert experience and knowledge, thus providing higher reliability and stability.

However, it also has the following disadvantages:

(a) Dependence on expert knowledge: Rule formulation depends on expert knowledge and experience and may be influenced by personal subjective factors.
(b) Limited applicability: Rule-based control strategies are usually applicable to specific scenarios and problems and may not perform well in complex systems and situations.

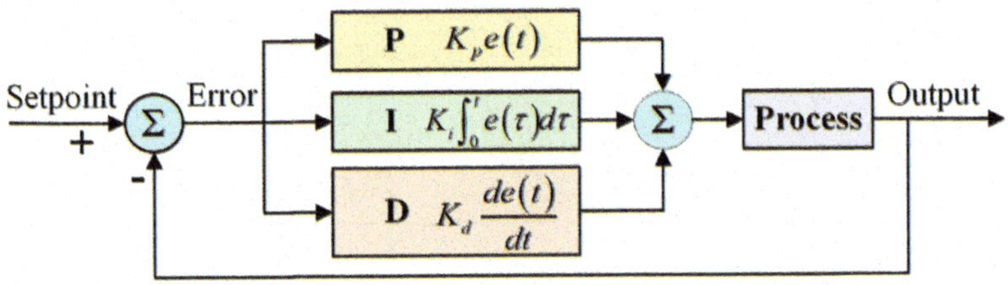

Fig. 3.21 Schematic of PID control

(c) Difficulty in dealing with unknown situations: When the system faces unknown situations or changes, rule-based control strategies may not be effective and require continuous updating and modification of rules.

Therefore, rule-based control methods perform well in some specific scenarios and problems, but may have limitations when facing complex and uncertain situations. Therefore, it is necessary to combine other control strategies and technologies to improve the adaptability and flexibility of the system.

(2) PID control method for tracking setpoint

The PID control method is a widely used technique in control systems to achieve desired system behavior. PID stands for proportional, integral, and derivative, which are three components that make up the control algorithm (Fig. 3.21).

The proportional component is responsible for providing an output proportional to the error between the desired setpoint and the actual system output. It adjusts the control effort based on the magnitude of the error, aiming to reduce it to zero. The proportional gain determines the sensitivity of the control response to the error, with a higher gain resulting in a stronger response.

The integral component integrates the error over time, which helps to eliminate steady-state errors. It continuously accumulates the error and adds it to the control output, making the control effort more accurate and precise. The integral gain determines the response to the accumulated error, with a higher gain leading to a faster correction of steady-state errors.

The derivative component calculates the rate of change of the error over time. It provides a prediction of the future error trend and helps to anticipate system behavior. By damping the control response, it reduces overshoot and oscillations. The derivative gain determines the response to the rate of change of the error, with a higher gain resulting in a stronger damping effect.

By combining these three components, the PID controller continuously adjusts the control output to minimize the error between the desired setpoint and the actual system response. The control algorithm calculates the control effort based on the weighted sum of the proportional, integral, and derivative terms. The gains of each component need to be properly tuned to achieve optimal control performance.

The PID control method provides a robust and versatile control strategy that can be applied to a wide range of systems, from simple to complex, to achieve desired system behavior and maintain stability.

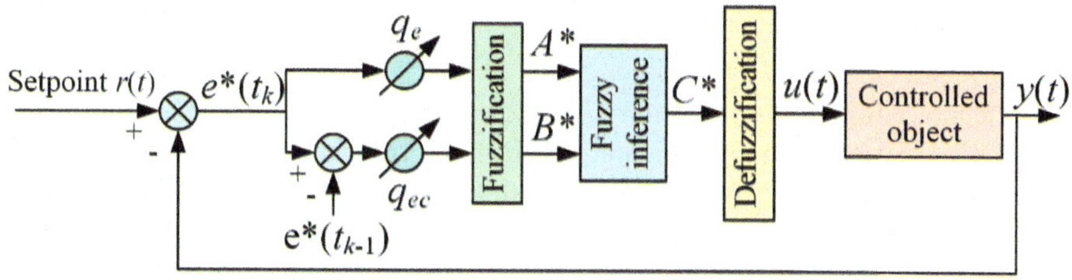

Fig. 3.22 Schematic of fuzzy control

(3) Fuzzy control method on tracking setpoints

Fuzzy control is a control method that utilizes fuzzy logic to handle complex and uncertain systems. It is based on the concept of fuzzy sets, which allow for the representation of vague or imprecise information (Fig. 3.22).

The basic principle of fuzzy control involves three main steps: fuzzification, rule evaluation, and defuzzification.

In the fuzzification step, the input variables of the system are converted into fuzzy sets. Fuzzy sets are defined by membership functions that assign a degree of membership to each element of the set. These membership functions represent the linguistic terms used to describe the system's inputs and outputs, such as "low", "medium", and "high".

The rule evaluation step involves applying a set of fuzzy rules to determine the appropriate control action. Fuzzy rules are typically expressed in the form of "IF–THEN" statements, where the "IF" part represents the fuzzy conditions based on the input variables, and the "THEN" part specifies the fuzzy actions based on the output variables. These rules capture the expert knowledge or experience about the system's behavior.

After the rule evaluation, the fuzzy outputs are obtained as fuzzy sets. These fuzzy outputs represent the control actions to be taken to regulate the system. However, they need to be converted into crisp values for implementation. This is done in the defuzzification step.

Defuzzification involves aggregating the fuzzy outputs and determining a single crisp value that represents the final control action. Various methods can be used for defuzzification, such as centroid, weighted average, or maximum membership.

The fuzzy control method allows for the handling of complex systems that are difficult to model mathematically or have uncertain dynamics. By using linguistic terms and fuzzy rules, it can capture the human-like decision-making process and effectively control the system's behavior.

Fuzzy control provides a flexible and intuitive approach to control systems, enabling the handling of uncertainty and imprecision in a wide range of applications.

3.5 Summary

This chapter mainly introduced the system configuration, models, and control loops of the HVAC water system, which are critical and fundamental to the system optimization and control. It can be found from this chapter that the pipe network, which connects the chiller system and user, is the most complex component in the HVAC water system. The main operating parameters of system like water temperature and flow will diffuse and propagate in the pipe network, and it is challenging to optimize the pipe network to keep hydraulic and thermal balance, which is critical to the indoor temperature control effect and user thermal comfort. This issue will be addressed in the following sections.

References

1. Tianyi Z, Liangdong M, Jili Z (2016) An optimal differential pressure reset strategy based on the most unfavorable thermodynamic loop on-line identification for a variable water flow air conditioning system. Energy Build 110:257–268. https://doi.org/10.1016/j.enbuild.2015.10.032
2. Stoecker WF (1975) Procedures for simulating the performance of components and systems for energy calculations, 3rd edn. United States
3. Fu XZ, Xiao YM (2018) Fluid network for transportation and distribution, 4th edn. China Architecture & Building Press, Beijing

Chapter 4
The Most Unfavorable Loop and the Thermodynamic Stability

Since the pipe network is the most critical component of HVAC water system, the optimization control for pipe network is becoming more significant. This chapter provides the definition and identification method of the most unfavorable thermodynamic loop (MUTL) under operating and regulating conditions. Furthermore, it introduces an innovative concept of pipe network stability known as thermodynamic stability, which aims to address stability issues arising from the operation and regulation of the HVAC water system. This concept is developed based on the characteristics of the HVAC water system and the control strategy of the terminal air-conditioning equipment.

4.1 The Most Unfavorable Loop in HVAC Water System

Figure 4.1 shows the water pressure diagram of an HVAC water system. There are n users in the system, and V_1 to V_n are regulating valves for each user. $p_0 - p_{m+1}$ are the pressure values of each node of the water system (m). E_1 *to* E_n are the load of each user, kW. S_1–S_n represent the impedance value of the branches, s^2/m^5. Q_{W1}–Q_{Wn} represent the water flow rate of each user (m^3/s). t_{CHWS1}–t_{CHWSn} represent the water supply temperature of each user (°C). t_{CHWR1}–t_{CHWRn} represent the return water temperature of each user (°C). The relationship between the available differential pressure (DP) and the water flow rate of the ith user ($i = 1, \ldots, n$) and the relationship between the load and the water flow rate can be expressed as Eqs. (4.1) and (4.2), respectively.

$$(p_j - p_{j+1}) = S_i Q_{Wi}^2, \tag{4.1}$$

$$E_i = \rho_W c_W Q_{Wi}(t_{CHWRi} - t_{CHWSi}). \tag{4.2}$$

T. Zhao et al., *Thermal Distribution Principle and Nonlinear Optimal Control for HVAC Water System*, Indoor Environment and Sustainable Building,
https://doi.org/10.1007/978-981-96-0377-0_4

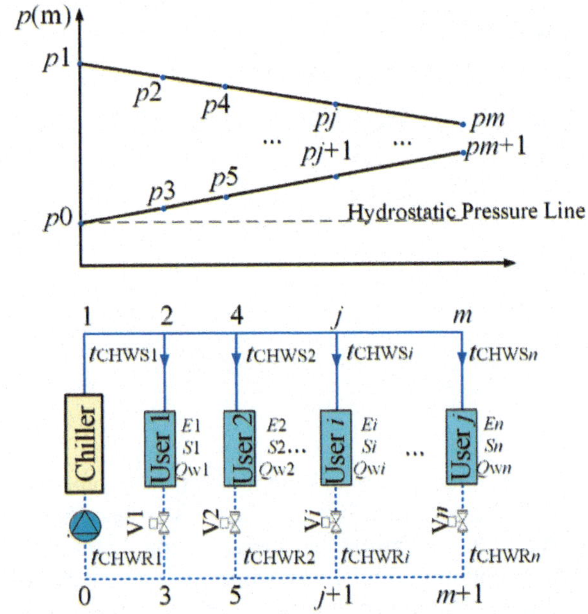

Fig. 4.1 Pressure diagram of air-conditioning water system

In the design stage of the HVAC water system, for the special building with large load in the inner area, the engineers first calculate the design load of each user and then determine the design flow rate of each user according to the temperature of chilled water supply and return under the design working condition (7 °C/12 °C) through Eq. (4.2). Subsequently, the design flow rate of the pump is calculated. Subsequently, the design flow rate of the pump is calculated. For ordinary air-conditioning buildings, the design load is determined by taking the maximum superposition value of the hourly load in each service room, and the pump's design flow rate can be determined according to it. By calculating the impedance value of each branch (with the control valve of each user branch fully open), the pressure distribution at each node is obtained using Eq. (4.1), thus obtaining the hydraulic parameter information of the water system under the design condition.

To ensure sufficient circulating power in the water system to satisfy all users' requirement of water flow rate under design conditions, it is usually necessary to select a loop with the largest loop resistance and determine the rated head of the circulating pump based on the required available differential pressure of the users on the loop under the design conditions. The loop with the greatest resistance is often called the most unfavorable loop, which is usually the loop of the user farthest away from the circulating pump. In Fig. 4.1, the most unfavorable loop is the loop of user n and the pump head must meet the available differential pressure required by user n, i.e., p_m–p_{m+1}.

It is evident that the most unfavorable loop mentioned above plays a crucial role in determining the pump model in the design stage of the HVAC water system. It is necessary to make a few remarks regarding the most unfavorable loops.

1. The most unfavorable loop has the greatest resistance compared with the other loops in the design condition. It is most unfavorable in the system resistance, and it is from the perspective of the hydraulic characteristics. To be precise, the most unfavorable loop discussed above should be referred to as the most unfavorable hydraulic loop (MUHL).
2. The MUHL is selected based on the design condition, where the flow rate of each user corresponds to the design flow rate. If there is a regulating valve in the branch, the valve position is set to be fully open.
3. Each pipe network has only one MUHL and its location is fixed.
4. Strictly speaking, the most unfavorable hydraulic loop is significant to the design stage of the pipe network, i.e., the pump head is determined according to its resistance. For the central air-conditioning system (especially, the large public building air-conditioning system), due to the diversified functions and personalized adjustment needs of the user, the system's operating parameters under actual conditions differ significantly from those under the design condition. In other words, the impact of the MUHL on system operation adjustments may not be apparent. To demonstrate this point, each parameter of Eqs. (4.1) and (4.2) will be analyzed.

Firstly, for Q_{Wi} in Eq. (4.2) under design conditions, Q_{Wi} is calculated according to the design load of each user and the design water supply and return temperature. However, under operating conditions, circumstances differ. Firstly, most of the times all users are in partial load conditions, resulting in the actual load value typically being less than the design load. Therefore, the variable E_i on the right side of Eq. (4.2) differs from design conditions. Secondly, the system's energy supply quality, which is reflected in the temperature of the chilled water supply (t_{CHWSi}), may also change. For example, the chilled water supply temperature of the system in actual operation could be 10 °C. Compared with its design value of 7 °C, the cold supply quality of the system decreases. Thirdly, the user-side's energy quality consumption also varies, which can be reflected by the return water temperature of chilled water, i.e., t_{CHWRi}. Taking the cooling coil as an example, the operating parameters of the wind side and the water side differ from the design conditions, and this change in the heat-exchange characteristics of the cooling coil makes the return water temperature inconsistent with the design value. Therefore, the Q_{Wi} for each branch under operating conditions will vary from the corresponding Q_{Wi} under design conditions. Due to these multiple factors, accurately predicting Q_{Wi} under operating conditions proves to be a challenge.

Secondly, for S_i in Eq. (4.1). Under design conditions, S_i is calculated on the basis of design flow, and the regulating valves are all fully open, so the corresponding S_i of each branch has little difference. The impedance of the connecting pipe sections in the MUHL is the largest among all the loops, so the total impedance of the MUHL is the largest. However, under operational conditions, the flow rate in each branch deviates from the design flow rate, leading to differing pipeline impedances from the design conditions. More critically, since the valve position in each branch fluctuates constantly, when the valve position is minimal, the Si in each branch may drastically differ. When the S_i difference across each loop surpasses the impedance difference of the corresponding connecting pipelines, the magnitude relationship of the total loop impedance will change. For example, under certain working condition, the valve position of the MUHL may be fully open, while the others are only slightly open. This could result in the impedance of these other loops surpassing that of the MUHL. Furthermore, in extreme instances, users on the MUHL may not require a flow supply; hence, its regulating valve would be closed, making the MUHL lose its reference value for system control.

Finally, due to the variability of S_i and $Q_{\mathrm{W}i}$ between the operating and design conditions of each branch, based on Eq. (4.1), the pressure differential demand for the users of each branch will alter under operational conditions. In certain scenarios, the pressure differential demand for users on the MUHL might be less than that of other users or even reach zero. Unlike under the design conditions, at this point, the differential pressure that the pump should provide should not be primarily determined by the pressure difference of the MUHL. Instead, the user with the highest pressure differential demand should be identified, and the working state of the pump should be reset accordingly.

During operation and control process, the MUHL is commonly adopted as referenced loop in the air-conditioning water system. For instance, the pump is usually adjusted based on the pressure difference of the user on the MUHL, where the differential pressure setpoint of the pump is usually determined according to the user's available pressure difference under the design condition. However, as the analysis above suggests, this adjustment mode could lead to several operational issues.

The most unfavorable loop in HVAC water system under operating conditions is no longer solely defined by hydraulic characteristics. It depends on multiple factors such as the dynamic characteristics of user load, the operating conditions of air-conditioning equipment, and the characteristics of the control loop. Hence, the differential pressure setpoint ($\mathrm{DP_{SPT}}$) needs to be adjusted in real time with reference to the load demand of the most unfavorable loop under operating conditions, which is called the most unfavorable thermodynamic loop (MUTL). Different types of HVAC water system have specific definition and identification methods for MUTL, which will be demonstrated in the next section.

4.2 Identification of the Most Unfavorable Loop

4.2.1 Identification of the MUTL in Continuous Regulated Water System

As mentioned in Sect. 3.1, based on the flow regulation mode of the terminal, there are two types of air-conditioning water system. Different types of HVAC water system have specific definition and identification methods for MUTL, which will be demonstrated in this section.

For the continuous regulated water system, based on the control loops mentioned in Sect. 3.3.2, the MUTL is defined as follows: In a HVAC water system with n branches, during the period of cooling, when the *i*th branch consistently operates under specific conditions during the cooling period, the loop containing this branch is defined as the most unfavorable thermodynamic loop (MUTL) for that period. These specific conditions include:

1. The valve position of branch is either fully open or nearly open, i.e., $\delta_{Vi} = 100\%$ or $\delta_{Vi} = \delta_{VMAXi}$.
2. For the variable air volume system, either the sending air temperature of AHU is higher than the setpoint or the return water temperature is higher than the setpoint, i.e., $t_{SAi} > t_{SASPT}$ or $t_{WRi} > t_{WRSPT}$. For the constant air volume system, either the return air temperature exceeds the setpoint, or the return water temperature surpasses the setpoint, i.e., $t_{RAi} > t_{RASPT}$ or $t_{WRi} > t_{WRSPT}$.

Some critical points about the MUTL include:

1. The MUTL can only manifest in the water system that simultaneously satisfies two conditions. Firstly, the air-conditioning equipment has an independent control loop that can regulate water volume according to the heat load. Secondly, the water pump is fitted with the variable frequency drive, allowing speed adjustments in response to changes in the system's hydraulic conditions.
2. The MUTL symbolizes a potential unfavorable condition in the operating process of the HVAC water system. Its presence indicates that the current working condition of the pump at least fails to meet the MUTL users' demands at the very least and the pump should be further regulated to enhance the circulation motivation.
3. During a period, there can be multiple MUTLs in the water system, and the position changes randomly. They could coincidentally be the MUHLs. The location is related to the change of the users' demand, the hydraulic characteristic of the pipe network, the thermal characteristic of the air-conditioning equipment, and the working characteristic of the actuator.
4. When the MUTL and MUHL are inconsistent, the pump's variable flow control strategy, which uses MUHL as a reference, becomes irrelevant. In such a scenario, the users' air-conditioning state fails to meet control requirements, and an alternative feasible strategy must be identified.

The MUTL is an important reference for the variable flow regulation. The real-time identification of the MUTL is essential for the efficient implementation of the variable flow control strategy. Then taking the continuously regulated water system in Fig. 3.1 as an example, the identification methods for several representative working conditions are outlined.

Working condition 1: There is only one MUTL in the system, and this loop is consistent with the MUHL, which is the loop that AHU9 is located in. Most of the time, δ_{V9} is the maximum value of each branch valve position, and the valve position change trend of each branch is basically the same. When δ_{V9} reaches 100%, the valve position of other branches may be closed to fully opened or remained between 50 and 80%. When the temperature of the room that AHU9 serves continues to rise, the valve position of other branches will not reach fully open. If this condition continues, it indicates that the load trend of the other users in the system is basically consistent and the user load for AHU9 is higher than the others. In this case, the DP_{SPT} strategy based on the MUHL is effective.

Working condition 2: There is only one MUTL in the system, and this loop is inconsistent with the MUHL. Assuming that AHU7 is located in the MUTL, the DP_{SPT} control strategy based on the MUHL will make the available differential pressure of AHU7 fail to meet the load requirement. In actual operation, the loop where AHU7 is located will remain as the MUTL for a long time, and the air-conditioning effect for users cannot be guaranteed.

Working condition 3: There are several MUTLs in the system, but they do not appear simultaneously. For example, the loops where AHU3, AHU4, and AHU5 are located all have potential to become the MUTL. During a certain period, the δ_{V3} is 100% and the duration that t_{SA3} remains above 17 °C reaches 5 min, assuming that the supply air temperature of the other AHUs is 16 °C.

At the same time, δ_{V4} and δ_{V5} are 60% and 70%, respectively, and t_{SA4} and t_{SA5} are 17 °C and 15.5 °C, respectively. Although the supply air temperature of AHU5 is already higher than its setpoint, the condition will not remain for a long time because V5 can still be increased to lower the temperature. Therefore, only AHU3 loop is the most unfavorable thermal loop at present.

Working condition 4: There are several MUHLs in the system simultaneously. For example, the loops where AHU3, AHU4, and AHU5 are located all have the possibility of becoming the MUTL. During a certain period, δ_{V3}, δ_{V4}, and δ_{V5} are all 100%, and t_{SA3}, t_{SA4}, and t_{SA5} are maintained at 17 °C, 18 °C, and 19 °C, respectively. At this time, the loops where AHU3, AHU4, and AHU5 are located are all the MUTLs, and the current available differential pressure of all these three loops fails to meet the requirement. In terms of the deviation of supply air temperature from the setpoint, the loop where AHU5 is located is the MUTL among those loops, and DP_{SPT} should be increased according to the user demand of AHU5 at this time.

Working condition 5: The valve positions of several branches in the system are fully open, but the supply air temperature of the AHU on some branches is not continuously higher than its setpoint. At a certain point, δ_{V2} and δ_{V7} are both 100%, and t_{SA2} and t_{SA7} are 18 °C and 17 °C, respectively. One minute later, δ_{V2} and δ_{V7} remain unchanged at 100%, but t_{SA2} has dropped to 15.8°C, while t_{SA7} has risen to

17.5 °C. If the identification period for the MUTL is one minute, the loop where AHU7 is located is the MUTL in the current period. Although δ_{V2} also maintains 100%, the loop where AHU2 is located is not the MUTL because the user load of AHU2 decreases during this period, and t_{SA2} does not consistently remain above the setpoint.

The five typical working conditions outlined above basically represent the identification process of the MUTL in continuous regulated variable flow water system. The pump should be regulated to avoid the emergence of the MUTL. However, if a MUTL arises within the water system, pump adjustments are necessary to ensure the available differential pressure for the user in the MUTL. Therefore, it is necessary to predict the potential location of the MUTL in advance before it materializes. Since the change in valve position of each branch is closely related to the identification of the MUHL and such changes occur more rapidly than alterations in supply air temperature, the MUTL can be identified in advance by the varying valve positions in each branch.

To help identify the MUTL, the definition of the maximum valve position loop is introduced. In a continuously regulated HVAC water system with n branches, if the valve position of *i*th branch reaches the maximum value among all branches in a certain time period, the loop containing this branch is located is defined as the maximum valve position loop of the HVAC water system for that period.

Compared with the other loops, the maximum valve position loop has the least flow adjustable margin, and it is the most potential candidate for the MUTL in the next period. The load fluctuation of each user makes the location of the maximum valve position loop mutable, resulting in multiple maximum valve position loops in the system simultaneously. Assuming that the valve position of the maximum valve position loop is δ_{MAX}, when δ_{MAX} approaches 100%, if the user's load on the maximum valve position loop continues to increase, this loop is prone to transfer into the MUTL. In order to avoid the occurrence of the MUTL or minimize its existence, DP_{SPT} should be proactively elevated. When δ_{MAX} is low (e.g., less than 50%), it indicates that the user in the maximum valve position loop still has the capacity to increase flow, reducing the likelihood of an ensuing MUTL. In such a case, the DP_{SPT} can be appropriately lowered to minimize the pump's energy consumption.

4.2.2 *Identification of the MUTL in On–Off Regulated Water System*

For the on–off regulated water system, except for the room temperature control loop mentioned in Sect. 3.3.2, there is another control loop, return water temperature control loop, which adjusts the opening of the continuous regulating valve on the *j*th riser, V*j*, to maintain the t_{WRj} as the setpoint depending on the deviation of return water temperature t_{WRj} from the setpoint.

Additional details regarding the room temperature control loop are provided below:

Firstly, the conventional on–off water regulation system does not include the room temperature control loop. On the one hand, by monitoring the return water temperature of each riser, the overall working condition of the corresponding FCUs can be understood to a certain extent. On the other hand, by controlling the return water temperature of each riser, the return water temperature of the system can be remained at a specific setpoint or fluctuate within a certain range, which contributes to the efficient operation of the chiller.

Secondly, since the number of terminals in the FCU system is large, the investment in the overall monitoring system for the FCUs is relatively high and its reliability is low. However, through loop of return water temperature, the working condition of a considerable number of FCUs can be reflected by the return water temperature of the riser, and those FCUs can be treated as an independent user in the continuous regulating water system. The return water temperature can be controlled by the regulating valve of the riser, reducing the need for an extensive monitoring system. Additionally, this approach enables effective monitoring of most FCUs' working conditions and control of the temperature difference between supply and return water in the system.

Based on the control loop of the return water temperature mentioned above, the definition of MUTL in the on–off regulating HVAC water system is given below. In an on–off regulating HVAC water system with n risers, as illustrated in Fig. 3.2, if the monitoring parameters corresponding to the *j*th risers continue to maintain the following two states during a certain period of cooling conditions, the loop where the *j*th riser is located is defined as the MUTL of the system for that period:

1. The continuous regulating valve position of the riser reaches fully open or close to fully open.
2. The return water temperature of the riser's chilled water consistently exceeds its setpoint.

Based on the room temperature control loop and aforementioned definition, the MUTL for the on–off regulating HVAC water system can be identified using a method similar to the continuous regulating water system. The method also utilizes the maximum valve position loop as a reference to realize the prediction of the MUTL.

4.3 Evaluation on Thermodynamic Stability

4.3.1 Definition of Hydraulic Stability

In the variable flow regulation of the HVAC water system, due to the constant adjustment of the pump and the valve opening of the equipment, the flow distribution of each branch in the network is of continually change and this kind of active flow change in one branch will cause the passive flow rate change in the other branches to a varying extent. This active flow change in one branch influences the passive flow rate change in other branches to varying degrees. If the change amplitude is too excessive, the system is regarded to be unstable. The concept of stability, in this context, relates to the hydraulic characteristics of the pipe network and is referred to as hydraulic stability.

As mentioned in Sect. 3.2.7, through the method of graph theory, the HVAC water system pipe network with any topology can be mathematically described by Eq. (4.3):

$$\begin{cases} B_r \cdot Q = 0 \\ B_r^T P_\mathrm{p} = I \cdot S \cdot Q \cdot |Q| + \Delta p - H_\mathrm{p} \end{cases}. \tag{4.3}$$

Q is a column vector of order m representing the flow rate of the branches. P_p is a column vector of order n representing the pressure at the nodes. I represents a unit matrix of order m. S is a column vector of order m, which represents the impedance of the branches. The dot product operator $(\cdot)$ is used. Δp is a column vector of order m representing the increase in pump head in the branches. H_p is a column vector of order m representing the pump head in the branches. Both the flow rate in each branch and the pressure at each node are function of the impedance (S).

The k branches that have to be examined out of m branches can be presented by U, which is a $k \times m$ matrix. As described in Eq. (3.29), the element of U is either 0 or 1, and each row of the matrix represents a branch and the position of 1 indicates the location of the branch in the system. If any branch is selected to be examined, then it is defined as active branch, denoted by C. The rest branches are the passive branches, denoted by U.

When the flow rate of the active branch C changes, it impacts the flow rate of the passive branches E accordingly. In an ideal scenario where control loops function optimally, any unintended flow rate changes in the passive branches can be counteracted by adjusting the pump speed or motorized valves in the branches. This adjustment in turn passively affects the flow rate of the active branch, denoted as $\Delta Q'_C$. The hydraulic stability, denoted by K_SC, is the proportion of the passive flow rate change in the active branch $\Delta Q'_C$ to the flow rate change in the active branch ΔQ_C, i.e., $K_\mathrm{SC} = \frac{\Delta Q'_C}{\Delta Q_C}$. A lower value of K_SC indicates higher stability in the active branches.

The calculation process of the K_SC is given below. Assuming that the impedance of the branch C can be regulated to adjust its flow rate, when the impedance changes by ΔS_C, ΔQ_C can be described by Eq. (4.4). Meanwhile, the change in flow rate of the passive branches E can be expressed by Eq. (4.5). To maintain the flow rate of branch E, the impedance of E should be adjusted to offset the passive change of the flow rate, as shown in Eq. (4.6).

$$\Delta Q_C = C \frac{\partial Q}{\partial S} C^T \Delta S_C, \tag{4.4}$$

$$\Delta Q_E = E\frac{\partial Q}{\partial S}C^T \Delta S_C, \tag{4.5}$$

$$\Delta Q_E' = E\frac{\partial Q}{\partial S}E^T \Delta S_E. \tag{4.6}$$

Since $\Delta Q_{\mathrm{E}}' = \Delta Q_{\mathrm{E}}$, the change in the impedance for passive branch E can be calculated by Eq. (4.7). The flow rate change in the active branch C caused by passive branch E can be presented as Eq. (4.8). Based on the definition of the hydraulic stability, combining Eqs. (4.4) and (4.8), K_{SC} can be rewritten by Eq. (4.9).

$$\Delta S_E = \left(E\frac{\partial Q}{\partial S}E^T\right)^{-1} E\frac{\partial Q}{\partial S}C^T \Delta S_C, \tag{4.7}$$

$$\Delta Q_C' = C\frac{\partial Q}{\partial S}E^T \Delta S_E = C\frac{\partial Q}{\partial S}E^T\left(E\frac{\partial Q}{\partial S}E^T\right)^{-1} E\frac{\partial Q}{\partial S}C^T \Delta S_C, \tag{4.8}$$

$$K_{\mathrm{SC}} = C\frac{\partial V}{\partial S}E^T\left(E\frac{\partial Q}{\partial S}E^T\right)^{-1} E\frac{\partial Q}{\partial S}C^T\left(C\frac{\partial Q}{\partial S}C^T\right)^{-1}. \tag{4.9}$$

Therefore, it can be found that the focus of the hydraulic stability lies in the active branch. It assesses the extent of the active branch flow rate affected by the other branches and the key parameter in the calculation is the flow rate.

4.3.2 *Definition of Thermodynamic Stability*

During the regulation process of variable flow HVAC water systems, there are active branches and passive branches within the system at any given time due to the random load fluctuation. The flow rates in the active branches should be regulated to satisfy the changing load demand, while the flow rates in the passive branches are not expected to change with relatively stable load demand. Take the water system in Fig. 3.1 as an example, the user load of the active branches increases, which indicates that the supply air temperature starts to exceed the setpoint. To maintain the setpoint of the supply air temperature, the valve opening in the active branch is adjusted to increase the flow rate. However, as the pump works under constant condition, the flow rates in the passive branches decrease, supplying lower cooling capacity and affecting the cooling for the users in passive branches.

In the case of stable user load, when the cooling capacity of the AHU is reduced by a certain level, the user air temperature (that is, the AHU return air temperature) begins to rise and the supply air volume increases, further elevating the AHU supply air temperature, leading to adjustments in the regulating valve of the passive branch. Such passive regulation is not driven by the changing user load demand but rather by the influence of active branch regulation on the passive branch flow. If this pattern

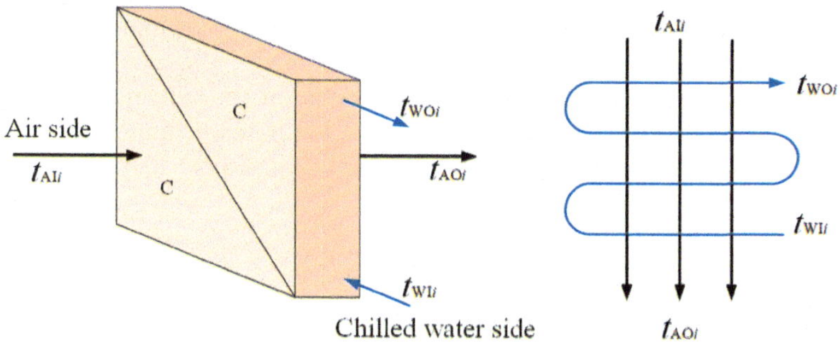

Fig. 4.2 Heat-exchange process in cooling coil (redrawn from Ref. [1], copyright 2013, with permission from ELSEVIER)

occurs in all passive branches, it not only weakens the hydraulic stability of the active branches but also increases the likelihood of frequent and random changes in the total impedance of the water system, negatively affecting the efficiency of DP control.

Therefore, it is necessary to discuss the impact of passive branch flow change on the cooling capacity of air-conditioning equipment in the above process.

Figure 4.2 shows the heat-exchange process of the AHU cooling coil in the ith passive branch (i.e., $i = 1, \ldots, e$, where e is the number of passive branches).

As shown in Fig. 4.2, t_{AIi} and t_{AOi} represent the air-side inlet and the outlet temperature (°C), respectively. t_{WIi} and t_{WOi} represent the chilled water-side inlet and outlet temperatures (°C), respectively. Q_{Ai} and Q_{Wi} represent the air-side and chilled water-side flow rates (m^5/h), respectively. The flow of air and chilled water in the cooling coil can be considered as a cross flow form with no mixing of fluids on both sides.

Based on the number-of-transfer-units method, Eq. (4.10) describes the heat transfer process of the cooling coil under dry condition.

$$E_{COILi} = \varepsilon C_{MIN}(t_{AIi} - t_{WIi}), \tag{4.10}$$

where E_{COILi} represents the cooling capacity of the cooling coil (kW), and C_{MIN} represents the minimum value of specific heat capacities of the airside and water-side fluid (kW/°C). Generally, C_{MIN} refers to the specific heat capacity of the air-side fluid, and ε represents the efficiency of the cooling coil, which can be calculated using Eq. (4.11):

$$\varepsilon = \frac{1 - e^{-NTU(1-C_R)}}{1 - C_R e^{-NTU(1-C_R)}}, \tag{4.11}$$

where NTU represents the number of transfer units of cooling coil, $NTU = \frac{kA}{C_{MIN}}$,

where k and A represent the heat transfer coefficient (kW/m^2 °C) and the heat transfer area (m^2), respectively. C_R is defined as $\frac{C_{MIN}}{C_{MAX}}$, where C_{MAX} denotes the maximum value of the specific heat capacities of the air-side and water-side fluid (kW/°C).

When the cooling coil operates under wet condition, the temperature involved in Eqs. (4.10) and (4.11) can be changed to the wet-bulb temperature for calculation. The air-side specific heat C_A can be calculated by Eq. (4.12).

$$c_{\mathrm{A}} = \frac{h_{\mathrm{AI}i} - h_{\mathrm{AO}_i}}{t_{\mathrm{AIWB}i} - t_{\mathrm{AOWB}i}}, \tag{4.12}$$

where $h_{\mathrm{AI}i}$ and $h_{\mathrm{AO}i}$ represent the air-side inlet and outlet air enthalpies (kJ/kg), respectively, and $t_{\mathrm{AOWB}i}$ is the air-side outlet air wet-bulb temperature (°C).

Then, the dry condition is analyzed. Considering the influence of ΔQ_{E} shown in Eq. (4.5) on the E_{COIL} shown in Eq. (4.10), the operating parameters of the cooling coil in the active branch i before and after regulation are distinguished by superscript (1) and (2), respectively, and the variation in the cooling capacity of the cooling coil $\Delta E_{\mathrm{COIL}i}$ can be expressed in Eq. (4.13).

$$\Delta E_{\mathrm{COIL}i} = \varepsilon^{(2)} C_{\mathrm{MIN}}^{(2)} \left(t_{\mathrm{AI}i}^{(2)} - t_{\mathrm{WI}i}^{(2)} \right) - \varepsilon^{(1)} C_{\mathrm{MIN}}^{(1)} \left(t_{\mathrm{AI}i}^{(1)} - t_{\mathrm{WI}i}^{(1)} \right). \tag{4.13}$$

The user load of the passive branches remain constant, which means the air-side inlet temperature and airflow are consistent before and after the active branch adjustment. Considering the C_{MIN} represents the specific heat capacity of the air-side fluid in most cases, C_{MIN} also remains unchanged before and after active branch adjustment. Meanwhile, the supply temperature of the cooling coil is assumed to be constant, then Eq. (4.13) can be rewritten as Eq. (4.14). This means that the change in the cooling capacity of the cooling coil depends on the difference in the efficiency of the cooling coil before and after the active branch adjustment.

$$\Delta E_{\mathrm{COIL}i} = C_{\mathrm{MIN}}^{(1)} \left(\varepsilon^{(2)} - \varepsilon^{(1)} \right) \left(t_{\mathrm{AI}i}^{(1)} - t_{\mathrm{WI}i}^{(1)} \right). \tag{4.14}$$

As shown in Eq. (4.11), the parameters related to ε include C_{MIN}, C_{MAX}, k, and A. After the active branch regulation, the flow rate change in the active branch $\Delta Q_{\mathrm{E}i}$ affects C_{MAX} and k, thus changing ε. The function of ε and the water flow rate $Q_{\mathrm{E}i}$ is denoted as $f_{\varepsilon - Qi}$, and then $\Delta E_{\mathrm{COIL}i}$ can be described below.

$$\Delta E_{\mathrm{COIL}i} = C_{\mathrm{MIN}}^{(1)} \left(t_{\mathrm{AI}i}^{(1)} - t_{\mathrm{WI}i}^{(1)} \right) \left[f_{\varepsilon - Qi} \left(Q_{Ei}^{(1)} + \Delta Q_{Ei} \right) - f_{\varepsilon - Qi} \left(Q_{Ei}^{(1)} \right) \right]. \tag{4.15}$$

Based on the analysis mentioned above, the definition of the thermodynamic stability is given below. For the continuously regulated variable flow water system, the stability of the passive branch can be defined by the impact of the flow regulation in the active branches on the cooling capacity of the air-conditioning equipment in the passive branches over a certain period, which is denoted by $\beta = [\beta_{ji}]$, where $j = 1, \ldots, c$ and $i = 1, \ldots, e$. Here, c and e represent the number of active branches and passive branches in the system, respectively, during the period. β_{ji} represents the impact of the flow regulation in the jth active branch on the cooling capacity of the ith passive branch's air-conditioning equipment over the given period.

$$\beta_{ji} = \frac{\Delta E_{\mathrm{COIL}i}}{\rho_W c_W \Delta Q_{Cj}\left(t_{\mathrm{CHWRC}j} - t_{\mathrm{CHWSC}j}\right)}, \tag{4.16}$$

where $\Delta E_{\mathrm{COIL}i}$ represents the change in the cooling capacity of the air-conditioning equipment in the ith passive branch (kW), as given in Eq. (4.15), and ΔQ_{Cj} represents the change in the flow rate of the jth active branch (m^3/s).

When $\beta_{ji} = 0$, it indicates that the flow regulation of the jth active branch has no effect on the ith passive branch. But in fact, it is unlikely that the active branch flow regulation has no effect on the cooling capacity of the passive branch air-conditioning equipment. Therefore, when β_{ji} is close to 0, the system can be considered stable.

When $\beta_{ji} < 0$, it indicates that the flow regulation of the jth active branch reduces the cooling supply of the air-conditioning equipment in the ith passive branch. The smaller the β_{ji}, the greater the impact.

When $\beta_{ji} > 0$, it indicates that the flow regulation of the active branch increases the cooling supply of the air-conditioning equipment in the ith passive branch. The greater the β_{ji}, the greater the impact.

4.3.3 Evaluation Index of Thermodynamic Stability

Definition of the thermodynamic stability in the previous section gives an intuitive representation of thermodynamic stability, but the β is difficult to be calculated by commonly used monitoring parameters. In addition, we are more concerned about the total magnitude of the corresponding cooling capacity change of each passive branch in the current time period, i.e., the overall thermodynamic stability, than the active branch that causes the change of the cooling capacity of the passive branch air-conditioning equipment. Therefore, it is necessary to seek a parameter that can reflect the overall thermodynamic stability of each passive branch in the current time period and can be obtained simply by monitoring parameters as an evaluation index of thermodynamic stability.

Therefore, this paper selects the maximum value of the absolute deviation of the air supply temperature of the air-conditioning equipment of each passive branch circuit from its setpoint in the time period under inspection as the evaluation index of the overall thermodynamic stability of each passive branch circuit, which is set as T_s, as in Eq. (4.17).

$$T_{Sj} = \overset{i=1}{\underset{m}{\mathrm{MAX}}}\left(\left|t_{\mathrm{SA}j}(i) - t_{\mathrm{SASPT}j}\right|\right), \tag{4.17}$$

where T_{Sj} represents the evaluation index of the thermodynamic stability (°C) for the *j*th passive branch. *i* represents the serial number of the sampling point during the inspection period. *m* is the total number of the sampling points during the inspection period. $t_{SAj}(i)$ denotes the *i*th sampling point of the supply air temperature of the air-conditioning unit on the passive branch *j* (°C). t_{SASPTj} denotes the supply air temperature setpoint of the air-conditioning unit on the passive branch *j* (°C).

It can be seen that the smaller the T_S is, the smaller the fluctuation of the cooling capacity of the passive branch air-conditioning equipment is, and the smaller the impact of the active branch flow regulation on the cooling effect of the passive branch users is. In other words, a smaller T_S leads to enhanced thermodynamic stability. For the control loop with air supply temperature as the control parameter, if the T_S of a passive branch is larger than the permissible fluctuation range of the setpoint of the air supply temperature of the AHU on the branch, the passive branch needs to carry out the relevant adjustment. The larger the T_S is, the larger the flow rate adjustment amplitude of the passive branch will be, and thus the larger the impact on the flow rate change of other branches will be, which will adversely affect the hydraulic stability of the system.

The reference value for the T_S depends on the control accuracy of the system and user load fluctuation. On the one hand, if the T_S reference value is too high, it will affect the user's air-conditioning effect or indoor environmental control accuracy. On the other hand, for users with frequent load fluctuation, the heat transfer conditions of air-conditioning equipment vary considerably over a given period of time, and the T_S reference value should not be too low. The relationship between T_S reference value and degree of thermodynamic stability is shown in Fig. 4.3.

In Fig. 4.3, the vertical axis parameter γ is a single-point fuzzy set [1, 2, 3], and the size of γ indicates the degree of thermodynamic stability. The smaller γ is, the better the thermodynamic stability is. $\gamma = 1, 2, 3$ corresponds to good, general, and poor thermodynamic stabilities, respectively. The horizontal axis indicates the reference value of T_S, and the T_S values corresponding to $\gamma = 1, 2, 3$ are T_{SLOW}, T_{SMID}, T_{SHIGH} (adjustable). If the system control accuracy required by the user is high, the reference value of the T_S is lower. If the system requires high robustness, i.e., by

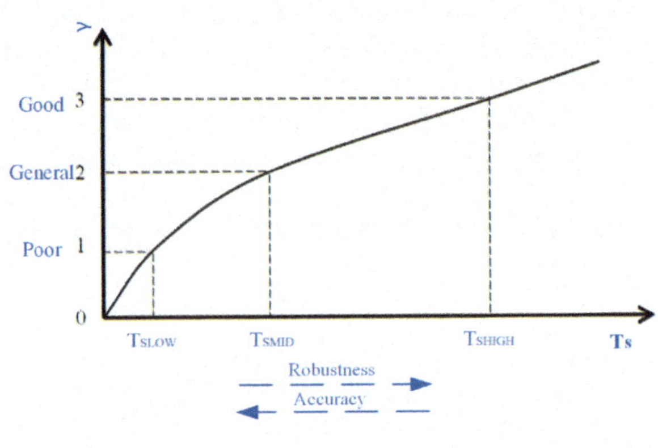

Fig. 4.3 Degree of thermodynamic stability versus *T*s

designing a larger range of air supply temperature robust area to make each passive branch corresponding to the condition parameter control loop which has strong anti-interference ability, the reference value of T_S can be increased appropriately. In this section, the reference values of T_{SLOW}, T_{SMID}, and T_{SHIGH} are 0.1 °C, 0.3 °C, and 0.5 °C, respectively.

The thermodynamic stability mentioned above is presented as an example of variable air volume air-conditioning equipment. For fixed air volume air-conditioning equipment, since the control valve position is adjusted according to the return air temperature, the key parameter for evaluation is return air temperature. Therefore, the parameter in the definition and evaluation index of thermodynamic stability should be changed from the supply air temperature to the return air temperature.

4.4 Summary

This chapter first points out that the traditional most unfavorable loop of HVAC water system is proposed based on the hydraulic characteristics, i.e., the most unfavorable hydraulic loop (MUHL). Then, the definition and identification method of the most unfavorable thermodynamic loop (MUTL) are proposed for two types of typical HVAC water system. Finally, in view of the stability problem of the pipe network, based on the study of hydraulic stability, combined with the heat transfer characteristics of the cooling coil, the concept and evaluation index of thermodynamic stability are proposed. The following conclusions can be drawn.

(1) The traditional concept of most unfavorable loop is the MUHL in the design stage to guarantee the circulating power demand of the pump.
(2) In the actual operation of the water system, the pump control strategy based on the MUHL is not ideal, and the reason is that the MUHL cannot reflect users' load demand.
(3) The MUHL can fully reflect the water supply and energy supply quality for the users, which exists objectively under the actual operating conditions of the system.
(4) The evaluation index of thermodynamic stability reflects the mutual influence of the hydraulic characteristics of each branch in the control loop of the terminal equipment and provides a new reference for optimizing the control strategy of the variable flow HVAC water system.

Reference

1. Tianyi Z, Jili Z, Liangdong M (2013) Experimental analysis of the thermal stability of the pressure control method for a variable flow air-conditioning water system. Build Environ 70:1–9. https://doi.org/10.1016/j.buildenv.2013.07.012

Chapter 5
Differential Pressure-Based Optimal Control Strategy

In this chapter, the differential pressure-based optimal control strategies will be discussed with reference to the definition and identification method of the MUTL including the optimization of DP and optimal control for chilled water pumps. Utilizing the simulation system, the feasibility of the differential pressure-based optimal control strategies are analyzed, taking the factors of the working condition parameter of the system, the energy conservation effect, and the thermodynamic stability into consideration.

5.1 Optimization of Differential Pressure

5.1.1 The DP Setpoint Optimal Reset Reference Variable

There are two requirements when determining the DP_{SPT}:

Firstly, the DP_{SPT} should satisfy the available differential pressure of all users in the system to avoid the occurrence of the MUTL. If the MUTL arises in the system, the DP_{SPT} should adjust promptly to meet the flow rate requirement for the MUTL.

Secondly, given the significant impact of the DP_{SPT} on energy consumption, the DP_{SPT} is expected to be as low as possible to minimize the system impedance while meeting the requirement of users' available differential pressure. This implies that the valve position of all branches should be in its maximum open state.

Based on the two points mentioned above, the maximum valve position loop is selected as the reference of the DP setpoint optimal reset. Taking the HVAC water system in Fig. 3.1 as example, assuming the loop where the *i*th branch is located is as the maximum valve position loop. Then for the maximum valve position loop, there is

T. Zhao et al., *Thermal Distribution Principle and Nonlinear Optimal Control for HVAC Water System*, Indoor Environment and Sustainable Building,
https://doi.org/10.1007/978-981-96-0377-0_5

$$\mathrm{DP} = (S_{\mathrm{V}i} + S_{\mathrm{B}i})Q_{\mathrm{W}i}^2 + \sum_{k=1}^{i}\left(S_{\mathrm{S}k}Q_{\mathrm{WS}k}^2 + S_{\mathrm{R}k}Q_{\mathrm{WR}k}^2\right). \tag{5.1}$$

The water flow in the ith branch can be demonstrated as:

$$\mathrm{DP} = (S_{\mathrm{V}i} + S_{\mathrm{B}i})Q_{\mathrm{W}i}^2 + \sum_{k=1}^{i}\left(S_{\mathrm{S}k}Q_{\mathrm{WS}k}^2 + S_{\mathrm{R}k}Q_{\mathrm{WR}k}^2\right). \tag{5.2}$$

In Eq. (5.2), since $S_{\mathrm{S}k}$, $S_{\mathrm{R}k}$, $Q_{\mathrm{WS}k}$, $Q_{\mathrm{WR}k}$, and $S_{\mathrm{B}i}$ change with working condition less significantly than $S_{\mathrm{V}i}$ and DP, to guarantee the users' requirement of $Q_{\mathrm{W}i}$, only DP and $S_{\mathrm{V}i}$ in Eq. (5.2) need to be adjusted. Therefore, for a given $Q_{\mathrm{W}i}$, there exist multiple working conditions characterized by different pairs of DP and $S_{\mathrm{V}i}$, where $S_{\mathrm{V}i}$ can be reflected by the valve position $\delta_{\mathrm{V}i}$.

If $\delta_{\mathrm{V}i}$ is less than 100% and the chilled water pump control can maintain DP as $\mathrm{DP_{SPT}}$, when the user load on the maximum valve position loop continues to increase, $\delta_{\mathrm{V}i}$ begins to increase to meet the demand of $Q_{\mathrm{W}i}$. When $\delta_{\mathrm{V}i}$ reaches 100%, Svi will be at the minimum and the regulator has no ability to increase the water flow rate any further. If the user load keeps increasing, the maximum valve position loop is likely to evolve into the MUTL.

In this case, DP in Eq. (5.2) should be increased by raising $\mathrm{DP_{SPT}}$, so that $Q_{\mathrm{W}i}$ can continue to increase with $\delta_{\mathrm{V}i}$ fully open until it satisfies the user's demand. When $\delta_{\mathrm{V}i}$ starts to be lower or $t_{\mathrm{SA}i}$ is no longer higher than the setpoint continuously, and there is no other MUTL in the system, the loop where branch i is located will return to the maximum valve position loop.

On the contrary, when the user load on the maximum valve position loop decreases, $\delta_{\mathrm{V}i}$ will decrease to reduce $Q_{\mathrm{W}i}$. To avoid excessive pump power consumption due to the high impedance of $\delta_{\mathrm{V}i}$, the DP value can be actively lowered by reducing $\mathrm{DP_{SPT}}$. With meeting the users' demand for flow rate $Q_{\mathrm{W}i}$, the $S_{\mathrm{V}i}$ should stay minimum, which means $\delta_{\mathrm{V}i}$ should be at a relatively high level.

According to the above analysis, the valve position $\delta_{\mathrm{V}i}$ in the maximum valve position loop can be used as the reference variable for $\mathrm{DP_{SPT}}$ optimal resetting.

5.1.2 *The DP Setpoint Optimal Reset Valve Position Domain*

In an actual control system, since the valve position changes frequently, in order to ensure the stability of pump control, the setpoint adjustment should not be too frequent. Therefore, $\mathrm{DP_{SPT}}$ cannot be adjusted strictly based on the exact value of $\delta_{\mathrm{V}i}$, but rather on a certain range of $\delta_{\mathrm{V}i}$, defined as the optimal setting field of valve position.

Figure 5.1 shows the relationship between resistance factor and valve position of equal percentage type valve. As the figure shows, when the valve position approaches fully open, the change rate of the regulating valve impedance decreases. In the shaded

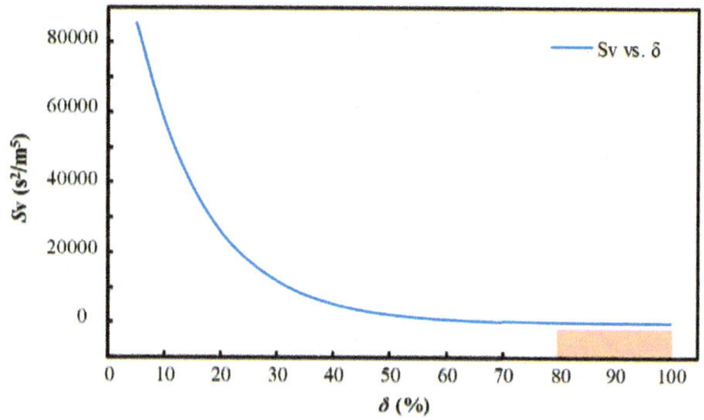

Fig. 5.1 Resistance factor versus valve position of equal percentage type valve (redrawn from Ref. [1], copyright 2016, with permission from ELSEVIER)

area (80% < δ < 100%) in Fig. 5.1, the change rate of the impedance is only 10 s^2/m^5 per percent of valve position, which means that the change of valve position slightly affects its impedance for the equal percentage regulating valve. In the other word, the change of valve position has no significant impact on the energy conservation of the pump within the shadow area's corresponding valve position domain which has a relatively wide area. Therefore, resetting the DP_{SPT} based on this valve position domain can reduce the adjustment frequency. The valve position range [δ_{MIN} (adjustable), 100%] that has the characteristics mentioned above is defined as the DP setpoint optimal reset valve position domain. For the regulating valve with other characteristics, the determination method of DP setpoint optimal reset valve position domain is similar.

5.1.3 The Best Valve Position Domain-Based DP Setpoint Optimal Reset Method

The regulation of the DP_{SPT} is applied to make the valve position of maximum valve position loop δ_{MAX} to stay in an ideal range. To predict the possibility of the occurrence of the MUTL, the upper limit should be close to 100%, and meanwhile, to ensure δ_{MAX} changes within the DP setpoint optimal reset valve position domain, the lower limit should be slightly higher than δ_{MIN}. Set the δ_{MAX} range [δ_{RSTMIN}, δ_{RSTMAX}] that satisfies all the above conditions as the DP setpoint optimal reset valve position domain. The aim of DP setpoint optimal reset is to maintain the δ_{MAX} within the optimal valve position domain.

(1) When $\delta_{MAX} > \delta_{RSTMAX}$, the maximum valve position loop tends to become the MUTL, and to avoid this situation, the DP_{SPT} should be increased. Accordingly, the available differential pressure for the users in the maximum valve position loop increases, lowering the δ_{MAX} into the optimal valve position domain.
(2) When $\delta_{MAX} < \delta_{RSTMAX}$, the valve position for the maximum valve position loop tends to be lower than the lower limit of the optimal valve position domain. In

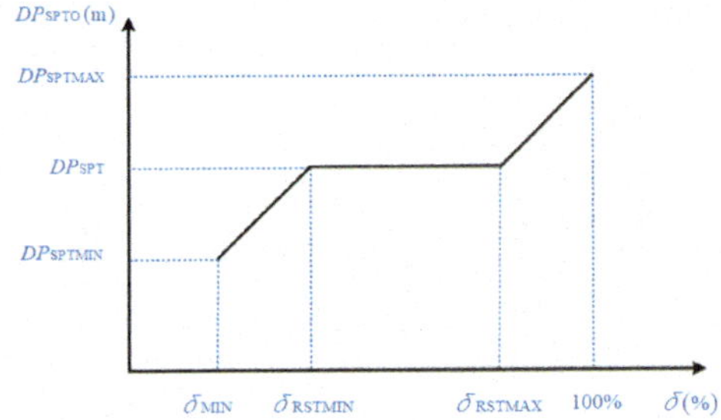

Fig. 5.2 Schematic diagram of reset strategy of differential pressure setpoint (redrawn from Ref. [1], copyright 2016, with permission from ELSEVIER)

order to prevent the valve position from decreasing and resulting in an increase in the system impedance, the DP_{SPT} needs to be reduced to raise δ_{MAX} within the optimal valve position domain.

(3) When $\delta_{MAX} \in [\delta_{RSTMIN}, \delta_{RSTMAX}]$, δ_{MAX} stays in an ideal range, which not only can avoid the occurrence of the MUTL, but also make the maximum valve position loop has relatively low impedance. Furthermore, the system operates under low impedance, which is the aim of DP setpoint optimal reset, and the DP_{SPT} at the moment is the ideal DP setpoint.

The DP setpoint optimal reset method can be illustrated by Fig. 5.2. The DP_{SPTMAX} and DP_{SPTMIN} represent the maximum and minimum value of the DP setpoint. The optimal DP_{SPT} can be represented by Eq. (5.3).

$$DP_{SPTO} = \begin{cases} (\delta - \delta_{MIN})\frac{DP_{SPT} - DP_{SPTMIN}}{\delta_{RSTMIN} - \delta_{MIN}} & \delta \in [\delta_{MIN}, \delta_{RSTMIN}) \\ DP_{SPT} & \delta \in [\delta_{RSTMIN}, \delta_{RSTMAX}] \\ (\delta - \delta_{RSTMAX})\frac{DP_{SPTMAX} - DP_{SPT}}{100\% - \delta_{RSTMAX}} & \delta \in (\delta_{RSTMAX}, 100\%] \end{cases}. \quad (5.3)$$

5.2 Analysis of Thermodynamic Stability

In this section, from the perspective of thermodynamic stability, the specific control cycle is analyzed, and the application effect of the DP_{SPT} adjustment strategy in terms of pipe network stability and pump energy saving is evaluated. Through comparison and analysis, the problems that should be paid attention to in the implementation of the optimal reset strategy are pointed out.

The schematic diagram of air-conditioning water system in simulation system is shown in Fig. 5.3. In the diagram, AHU1–AHU6 are variable air volume air-conditioning units. V1–V6 are continuous regulating valves. G is the sensor for the total chilled water volume, and the ΔP is the water flow DP sensor. Each node and each pipe are numbered, and the diameter for the pipes are given in the diagram. The system has a building energy management system (BEMS) that monitors the

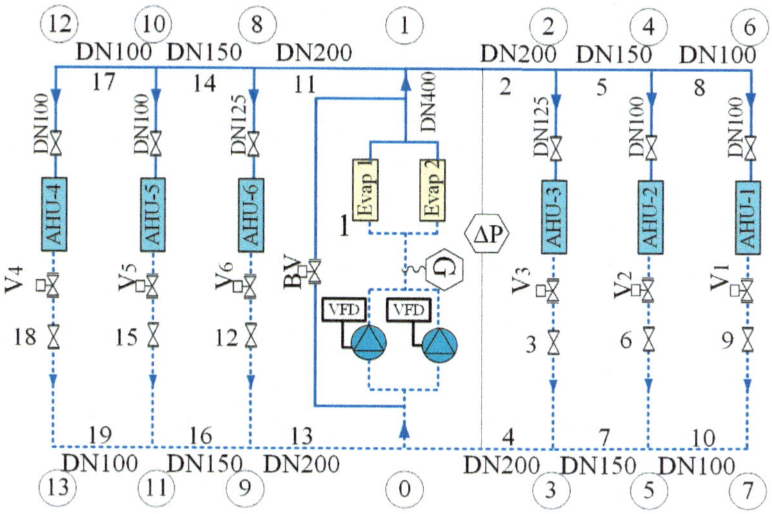

Fig. 5.3 Schematic diagram of air-conditioning water system in simulation system (redrawn from Ref. [1], copyright 2016, with permission from ELSEVIER)

operational parameters in real time, and the past operational data can be obtained in this system.

5.2.1 The Initial Parameter for the Simulation

The working condition parameters used in this model is selected from the operating data of an actual project, as shown in Table 5.1.

The 12th branch where AHU6 is located is selected as the active branch, and AHU1–AHU5 are taken as the passive branch. The full stroke cycle for the valve actuator in the active branch is 188 s, which is taken as the examined period, and

Table 5.1 Initial simulation parameters (data from Ref. [2], copyright 2013, with permission from ELSEVIER)

Simulation Parameter	AHU1	AHU2	AHU3	AHU4	AHU5	AHU6
Inlet water temperature of each AHU t_{WI} (°C)	7					
Inlet temperature of each AHU t_{AI} (°C)	28					
Enthalpy of inlet air h_{AI} (kJ/kg)	63.21					
Outlet temperature of each AHU t_{AO} (°C)	15					
AHU supply air volume Q_{SA} (m^3/s)	20.77	20.77	19.52	18.87	18.87	26.07
Valve opening δ (%)	62	42	16	17	15	95

Table 5.2 Evaluation parameters (data from Ref. [2], copyright 2013, with permission from ELSEVIER)

Evaluation parameter	Unit	Description
DP	m	Reference DP
P_P	kW	Pump electric power
t_{AO}	°C	Outlet temperature of each AHU
$K_{SC}(C, E)$		The evaluation index of hydraulic stability
T_S	°C	The evaluation index of thermodynamic stability

the sampling period is 2 s. During the simulation process, the regulating valve is assumed to be still at the beginning, and then, the regulating valve will act based on the simulation result. The evaluation parameters are listed in Table 5.2.

5.2.2 *Experimental Control Strategy*

The structural equation of pipe network structure and the matrices for the investigated branch matrix C and E are established. The 13-order row vector is defined as $\text{DP}_m = (1, 0, \ldots, 0)$. The position of 1 represented the node position of the constant pressure point in the pipe network. The simulation strategies are listed in the following.

Control Strategy 1: During the regulating period, the pump control cycle is longer than the regulating valve acting cycle, and the differential pressure is out of control. Since the pump speed remains stable during the regulating period, the change in DP (ΔDP) and the flow rate change in passive branch (ΔQ_E) only depend on the valve position change in the active branch. ΔDP and ΔQ_E can be expressed by Eq. (5.4).

$$\begin{cases} \Delta \text{DP} = \text{DP}_m \frac{\partial P}{\partial S} C^T \Delta S_C \\ \Delta Q_E = E \frac{\partial Q}{\partial S} C^T \Delta S_C \end{cases}. \tag{5.4}$$

Control Strategy 2: During the regulating period, the pump control cycle is shorter than the regulating valve acting cycle, and the differential pressure setpoint remains at the initial state. ΔDP and ΔQ_E depend on the valve position in the active branch and the pump speed change. The DP control strategy can be presented as follows.

$$\text{DP}_m \frac{\partial P}{\partial S} C^{\text{T}} \Delta S_C = -\text{DP}_m \frac{\partial P}{\partial w} C^{\text{T}} \Delta w. \tag{5.5}$$

In Eq. (5.5), Δw represents the required water pump speed to keep DP_{SPT}. After Δw is determined, Q_E can be calculated as follows.

$$Q_E = E \frac{\partial Q}{\partial S} C^{\text{T}} \Delta S_C + E \frac{\partial Q}{\partial w} C^{\text{T}} \Delta w. \tag{5.6}$$

Control Strategy 3: During the regulating period, the pump control cycle is shorter than the regulating valve acting cycle, and DP_{SPT} is reset based on the optimal control strategy. In order to intuitively use hydraulic parameters for analysis, so as to combine with changes in hydraulic stability, DP_{SPT} is adjusted according to the active branch flow rate change. When the flow rate decreases, DP_{SPT} decreases, and vice versa. The adjustment amplitude of DP_{SPT} should be determined according to the characteristics of the regulating valve and the hydraulic characteristics of the pipe network. To facilitate the comparison, the hydraulic condition of the constant pressure difference strategy corresponding to Strategy 2 is taken as a reference, corresponding to sampling time *i*, and the flow value of the active branch in Strategy 2 is set as Q_{Ci} and the adjustment amplitude of DP_{SPT} can be expressed by Eq. (5.7).

$$\mathrm{DP_{SPT}} = \left(Q_{C(i-1)} - Q_{C(i-2)}\right) + \lambda\left[\left(Q_{Ci} - Q_{C(i-1)}\right) - \left(Q_{C(i-1)} - Q_{C(i-2)}\right)\right]. \quad (5.7)$$

where $Q_{C(i-1)}$ represents the active branch flow rate at the previous time point, m^3/s. $Q_{C(i-2)}$ represents the active branch flow rate at the most recent two time point, and m^3/s. λ is the DP setpoint adjustment constant.

It can be seen that the λ is positively correlated with the DP_{SPT} adjustment amplitude. In order to explore the influence of DP_{SPT} adjustment amplitude on each evaluation parameter, λ was simulated with 50, 100, 200, 400, 800, and 1200, and corresponding strategies were 3_1, 3_2, 3_3, 3_4, 3_5, and 3_6, respectively. The control equation of pressure difference is

$$\Delta \mathrm{DP_{SPT}} = \mathrm{DP}_m \frac{\partial P}{\partial S} C^{\mathrm{T}} \Delta S_C + \mathrm{DP}_m \frac{\partial P}{\partial w} C^{\mathrm{T}} \Delta w. \quad (5.8)$$

According to Eq. (5.5), Δw can be determined, and using the result in Eq. (5.4), the ΔQ_E can be determined.

In Eqs. (5.4)–(5.8), the gain matrix expression of impedance, speed, node pressure, and branch flow can be derived from the pipe network structural equation and calculated by Q, S, and ΔH at the sampling time. The expression of each gain matrix is presented below.

$$\frac{\partial Q}{\partial S} = M^{-1}(A^T(AM^{-1}A^T)^{-1}AM^{-1} - I)I|Q|Q, \quad (5.9)$$

$$\frac{\partial Q}{\partial w} = -M^{-1}A^T(AM^{-1}A^T)^{-1}AM^{-1}\frac{\partial \Delta H}{\partial w} + M^{-1}\frac{\partial \Delta H}{\partial w}, \quad (5.10)$$

$$\frac{\partial P}{\partial S} = (AM^{-1}A^T)^{-1}AM^{-1}I|G|G, \quad (5.11)$$

$$\frac{\partial P}{\partial w} = -(AM^{-1}A^T)^{-1}AM^{-1}\frac{\partial \Delta H}{\partial w}. \quad (5.12)$$

In Eqs. (5.9)–(5.12), $M = 2IS|Q| - \frac{\partial \Delta H}{\partial Q}$.

After obtaining ΔQ_E of each strategy at each sampling time, the thermodynamic stability evaluation indicator, T_S, was subsequently calculated.

5.2.3 Analysis of Experimental Results

As shown in Fig. 5.4, under the condition that the pump is not controlled (Strategy 1), the pump flow decreases as the valve position of the active branch decreases, leading to higher head and pressure difference, and vice versa. When adopting Strategy 2, the pump is automatically adjusted. When the valve position of the active branch decreases, the pump speed will be reduced to maintain the initial pressure difference, and vice versa. With Strategy 3, the DP_{SPT} is reset, reducing the pressure difference and pump power consumption.

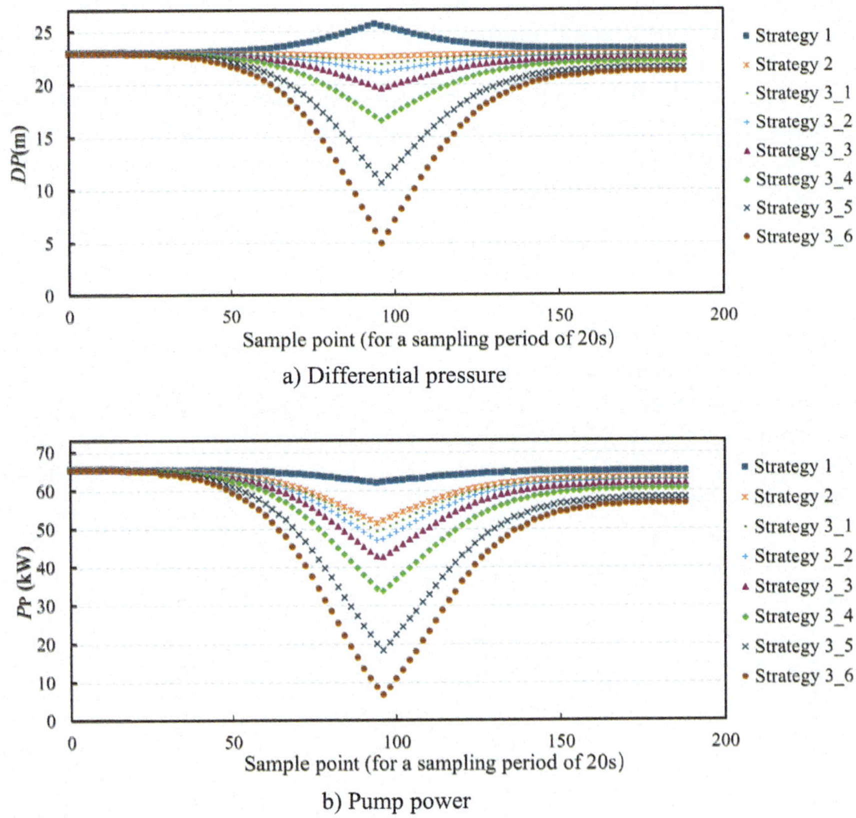

Fig. 5.4 Experimental results for differential pressure and hydraulic pump power (redrawn from Ref. [2], copyright 2013, with permission from ELSEVIER)

Figure 5.5 shows the simulation results of the supply air temperature of each unit during the regulation process. According to the variation of supply air temperature of passive branch units AHU1 to AHU5, as shown in Fig. 5.5a–e, it is under the control of constant pressure difference (Strategy 2) that each passive branch has the best thermodynamic stability, and the supply air temperature of each unit remains stable. When the pump is without control, the valve opening of the active branch reduces, causing the increasing passive branch flow, which enhances the cooling capacity of the cooling coil and lowers the supply air temperature, and vice versa.

For Strategy 3, since the DP_{SPT} is adjusted according to the flow required by the active branch, when the valve position is reduced, the DP_{SPT} also decreases accordingly, thus inhibiting the increasing trend of passive branch flow rate affected by the active branch to a certain extent. When the change amplitude of the setpoint increases, the flow rate of the passive branch decreases conversely compared with the initial value, and vice versa. For Strategy 3_5 and Strategy 3_6, although they have significant pump energy-saving effect, the maximum fluctuation range of supply air temperature of the passive branch unit has exceeded 0.5 °C due to the large change step of the DP_{SPT}. This indicates that pump energy saving is achieved at the expense of thermodynamic stability.

Figure 5.5f shows the variation of supply air temperature of AHU6 active branch unit. In the simulation process, since the supply air volume and inlet air temperature of AHU6 remain constant, the variation range of supply air temperature reflects the cooling capacity adjustability of the unit. As the figure shows, Strategy 3 improves the adjustability of the active branch. The larger the DP_{SPT} change step is, the better the adjustability is, and the shorter the adjustment time the unit needs to meet the user's cooling capacity demand is. In summary, the adjustment step of the DP_{SPT} should balance the energy consumption, thermodynamic stability, and adjustability of the pump.

Table 5.3 gives the average values of the evaluation indexes over the regulation period. Based on the data in the table, the following observations can be made:

(1) The active control of the pump can improve the hydraulic stability. With the increase of the adjustment constant λ of DP_{SPT}, the hydraulic stability has a certain degree of improvement, and the power consumption of the pump significantly reduces.

(2) For the dual-loop water system, as shown in Fig. 5.3, when the water pump is not controlled, compared with AHU1–AHU3, AHU4 and AHU5 have low thermodynamic stability because they are in the same hydraulic loop with AHU6 which is located in the active branch. Controlling DP can effectively improve the total thermodynamic stability of all the branches. For AHU1, AHU2, and AHU3 in the other loop, after adopting the Strategy 2, all the T_S become 0 and the thermodynamic stability is high, which is also the advantage for the dual-loop water system. For the Strategy 3, with the increase of λ, the thermodynamic stability between two loops tends to be balanced. When λ increases to 100, the T_S for all 5 units is 0.08. If λ continues to increase, the T_S for AHU1–AHU3 becomes larger than AHU4 and AHU5. This indicates that compared with the

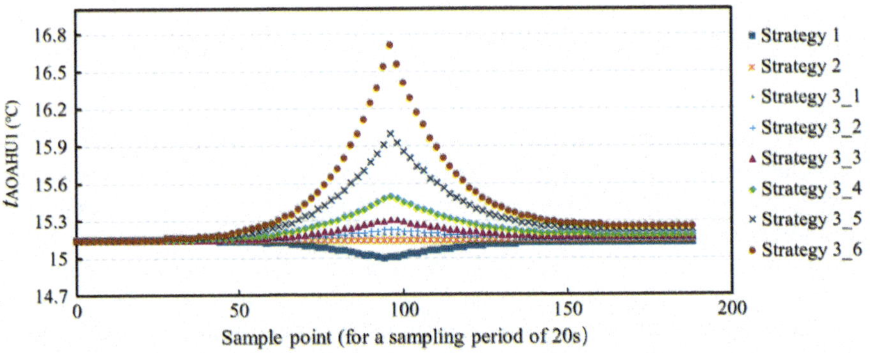

a) Supply air temperature for AHU1

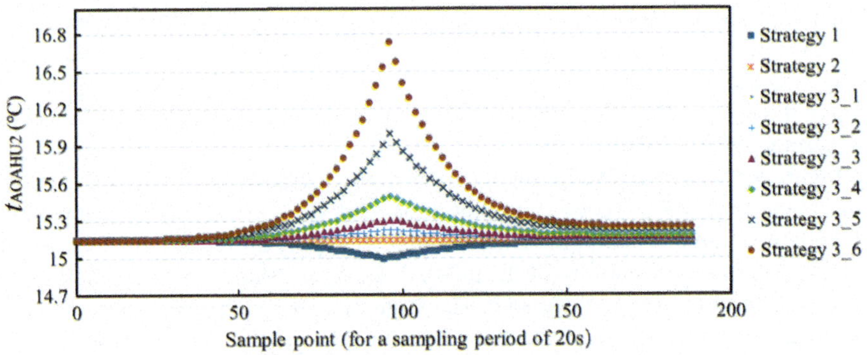

b) Supply air temperature for AHU2

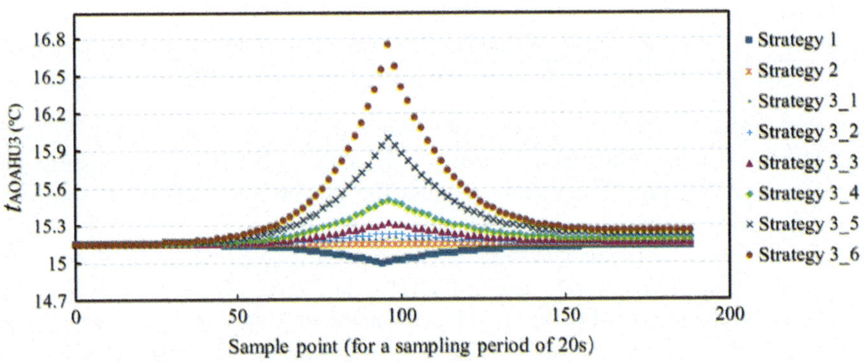

c) Supply air temperature for AHU3

Fig. 5.5 Changes in supply air temperatures for various branch units (redrawn from Ref. [2], copyright 2013, with permission from ELSEVIER)

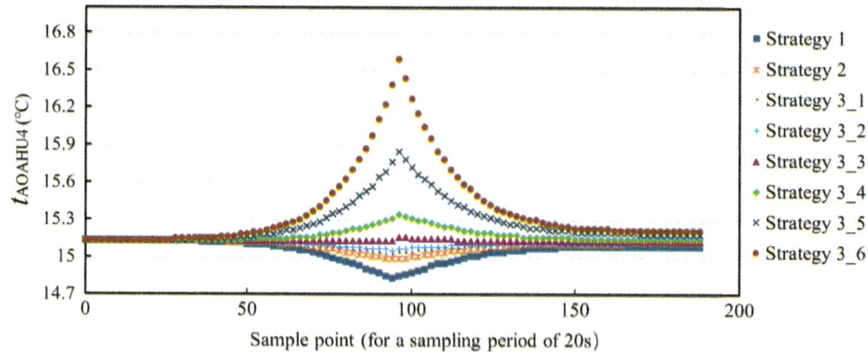

d) Supply air temperature for AHU4

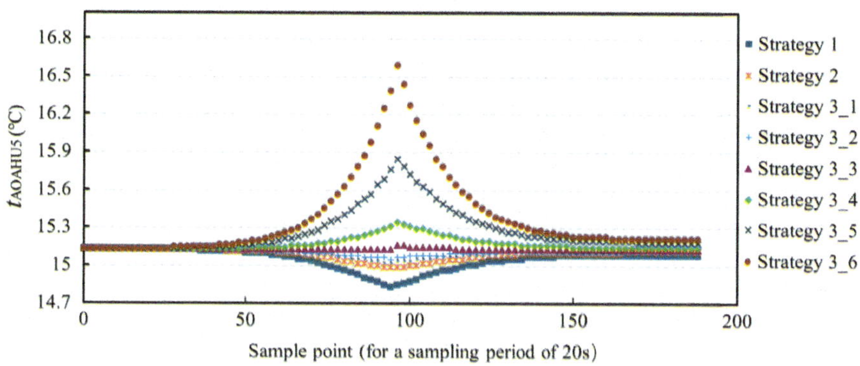

e) Supply air temperature for AHU5

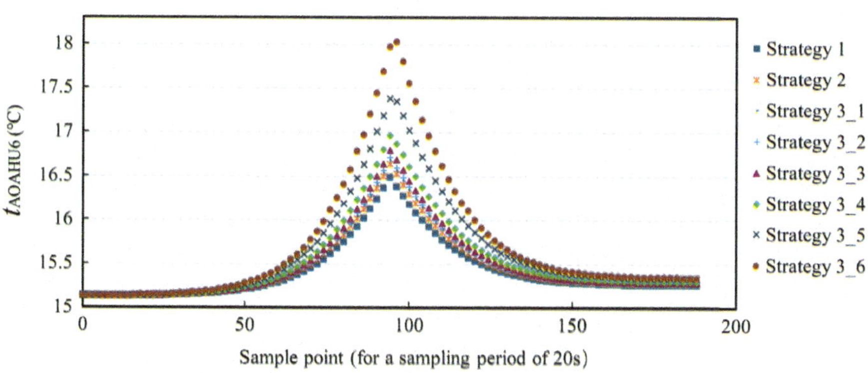

f) Supply air temperature for AHU6

Fig. 5.5 (continued)

Table 5.3 Average of evaluation parameters (redrawn from Ref. [2], copyright 2013, with permission from ELSEVIER)

Evaluation parameter	1	2	3_1	3_2	3_3	3_4	3_5	3_6
T_{SAHU1} (°C)	0.15	0	0.04	0.08	0.16	0.35	0.85	1.57
T_{SAHU2} (°C)	0.15	0	0.04	0.08	0.16	0.35	0.85	1.59
T_{SAHU3} (°C)	0.16	0	0.04	0.08	0.17	0.35	0.85	1.60
T_{SAHU4} (°C)	0.30	0.14	0.1	0.08	0.03	0.22	0.71	1.46
T_{SAHU5} (°C)	0.30	0.14	0.1	0.08	0.03	0.22	0.71	1.46
$K_{SC}(C,E)$	0.116	0.114	0.113	0.113	0.112	0.110	0.106	0.103
DP (m)	23.68	22.91	22.72	22.52	22.14	21.37	19.89	18.57
P_P (kW)	64.6	61.8	61.1	60.4	59.0	56.4	51.7	47.8

active branch regulation, the reduction magnitude of the DP_{SPT} causes stronger impact on the thermodynamic characteristic of AHU1–AHU3.

(3) For a certain regulation period, the good hydraulic stability cannot represent low fluctuation of cooling capacity. As Table 5.3 shows, although the Ksc decreases with the increase of λ, the T_S increases steadily. When λ exceeds a certain threshold, the thermodynamic stability significantly worsens. For instance, when adopting Strategy 3_2, the T_S for all the branches is less than 0.1 °C, which means that the system is of high thermodynamic stability. When λ increases to 200, the maximum T_S for all branches is 0.17 and the thermodynamic stability is still good. When λ increases to 400, the T_S for AHU3 climbs to 0.35, and the thermodynamic stability is relatively low. If λ keeps increasing, the thermodynamic stability will deteriorate further. For example, the T_{SAHU1}, T_{SAHU2}, and T_{SAHU3} corresponding to Strategy 3_5 exceed 0.5 °C and the regulating valve will regulate passively to a large degree, thus affecting the impedance of the pipe network and the flow distribution and causing new unstable factors.

(4) The DP_{SPT} optimal reset strategy (Strategy 3) has significant advantages on the energy conservation and can effectively improve or balance the thermodynamic stability of the passive branches when λ varies within a certain range. As the table shows, selecting λ within 200 and 400, making the strategy between Strategy 3_3 and Strategy 3_4 (as the shaded area in the table shows), then the ideal thermodynamic stability and effective pump energy conservation can be achieved.

Table 5.4 Setting parameters and evaluation parameters (data from Ref. [1], copyright 2016, with permission from ELSEVIER)

Predefined parameter	Predefined value	Evaluation parameter	Reference value
The upper limit of the best valve position domain δ_{RSTMAX} (%)	98	DP setpoint DP_{SPT} (m)	55
The lower limit of the best valve position domain δ_{RSTMIN} (%)	92	Total water pump power P_{PT} (kW)	78
AHU surface cooler supply air temperature t_{AO} (°C)	16	Water pump rotating speed ratio w	1
Chilled water supply temperature t_{CHWS} (°C)	7	Water pump efficiency η_P (%)	
		Each regulating valve position's maximum value δ_{MAX} (%)	100
		Each regulating valve position's average value δ_{AVE} (%)	100
		Chilled water return water temperature t_{CHWR} (°C)	

5.3 Simulation of the Optimal Control Strategy

5.3.1 The Simulation Parameter and Procedure

In this section, the optimal control strategy mentioned in Sect. 5.1 and several typical control strategies are applied to the simulation model.

The simulation conditions were selected from 364 groups of hourly conditions in the system. Simulation setting parameters and evaluation parameters are shown in Table 5.4. To facilitate comparison of simulation results of various strategies, all evaluation parameters in Table 5.4, except the chilled water return temperature and pump efficiency, are presented in percentage form (100% × simulation result/reference value). The rated working condition parameter of the pump is taken as the reference value of pump performance parameter, and the rated pump head is taken as the reference value of the DP_{SPT}.

5.3.2 The Control Strategy for Simulation

To compare the effect of the optimal resetting strategy and the other strategies, the following five representative DP_{SPT} resetting strategies are adopted in the simulation.

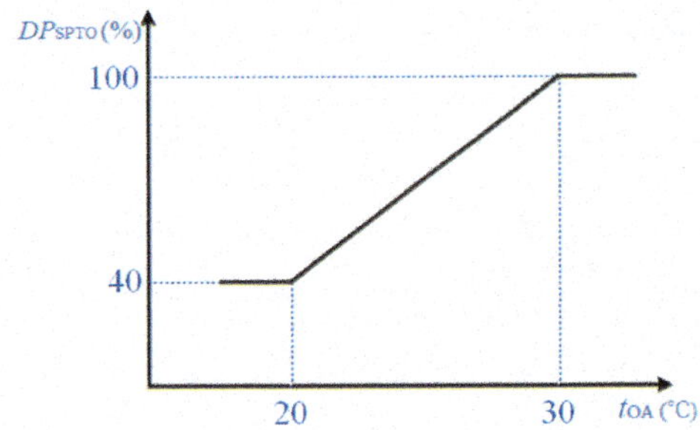

Fig. 5.6 Optimal reset strategy of DP_{SPT} based on outdoor temperature (redrawn from Ref. [1], copyright 2016, with permission from ELSEVIER)

Strategy 1: constant flow at the chiller side and a constant DP_{SPT} at the user side. The controller adjusts the opening of bypass valve to maintain the DP of the supply and return water main pipes as its setpoint.

Strategy 2: variable flow at the chiller side and a constant DP_{SPT} at the user side. The controller adjusts the primary water pump speed to maintain the DP as the setpoint.

Strategy 3: The DP_{SPT} is reset based on the outdoor temperature (t_{OA}), which is reckoned to indicate users' load demand change (Fig. 5.6).

Strategy 4: DP_{SPT} is reset based on the return chilled water temperature t_{CHWR}. When the load is reduced, both the supply air volume Q_{SA} and the inlet air temperature t_{AI} of the cooling coil decrease. In order to maintain the supply air temperature as setpoint, the valve will be adjusted to change the flow rate, leading to a change in the outlet water temperature of the cooling coil, thereby altering t_{CHWR}.

Figure 5.7 illustrates the relationship between the outlet water temperature t_{WO} of the cooling coil of AHU6 and Q_{SA} and t_{AI}. The figure demonstrates that t_{WO} generally increases as Q_{SA} and t_{AI} decrease, while the change of t_{CHWR} mainly depends on the t_{WO} changes in the AHUs. If t_{CHWR} is used as an indicator of load changes, it can be inferred that the t_{CHWR} drops with an increase in load, and vice versa. The control strategy is shown in Fig. 5.8.

Strategy 5: The DP_{SPT} is adjusted based on the valve position of AHU1 δ_1 in MUHL as reference, ensuring δ_1 within the best valve position domain [92%, 98%].

5.3.3 *The Simulation Result Analysis*

Figure 5.9 shows the simulation results.

Table 5.5 lists the average values of each evaluation parameter in 364 groups of simulation conditions, and the data in the table can generally reflect the application effect of each strategy.

Based on the above results, there are some points worth analyzing.

Firstly, the DP_{SPT} of the optimal reset strategy and Strategy 5 is relatively lower compared with the other four strategies. Accordingly, the total power for the pump is also low. As the valve position in Table 5.5 shown, the optimal resetting strategy

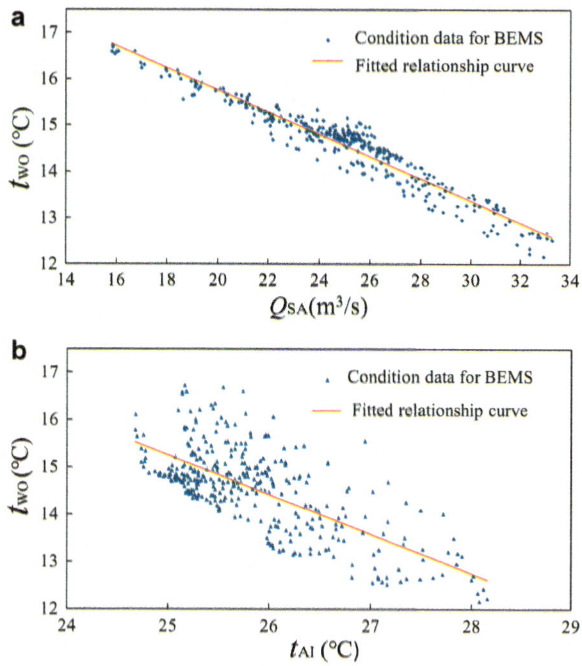

Fig. 5.7 AHU6 cooling coil chilled water outlet temperature and supply air volume versus inlet air temperature: **a** effect of supply air volume on chilled water outlet temperature, and **b** effect of inlet air temperature on chilled water outlet temperature (redrawn from Ref. [1], copyright 2016, with permission from ELSEVIER)

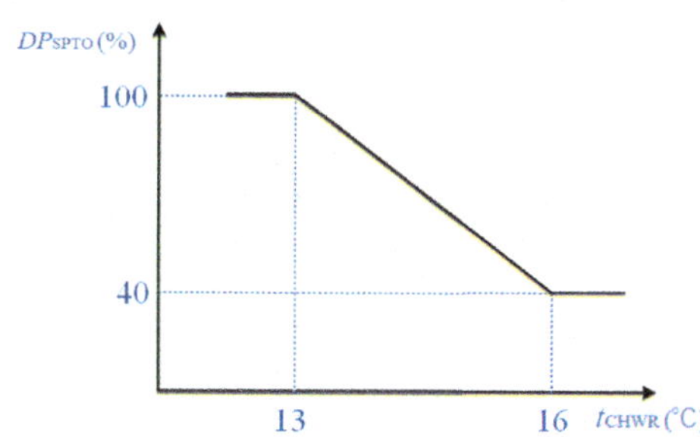

Fig. 5.8 Optimal reset strategy of DP_{SPT} based on return chilled water temperature (redrawn from Ref. [1], copyright 2016, with permission from ELSEVIER)

makes the valve position in each branch remaining within the best valve position domain in most conditions, thus decreasing the pump energy consumption on the regulating valve. The P_{PT} average value for the resetting control strategy is 22.61% of Strategy 1, 31.67% of Strategy 2, 42.36% of Strategy 3, and 53.14% of Strategy 4.

Secondly, the average chilled water return temperature of the system under the constant flow operation strategy on the cold source side (Strategy 1) and the variable flow operation strategy on the cold source side (Strategy 2–5 and optimal reset strategy) are 13 °C and 15.47 °C, respectively, which means that the unit evaporation temperature corresponding to the variable flow strategy on the cold source side is higher. This is conducive to improving the unit performance coefficient. On the

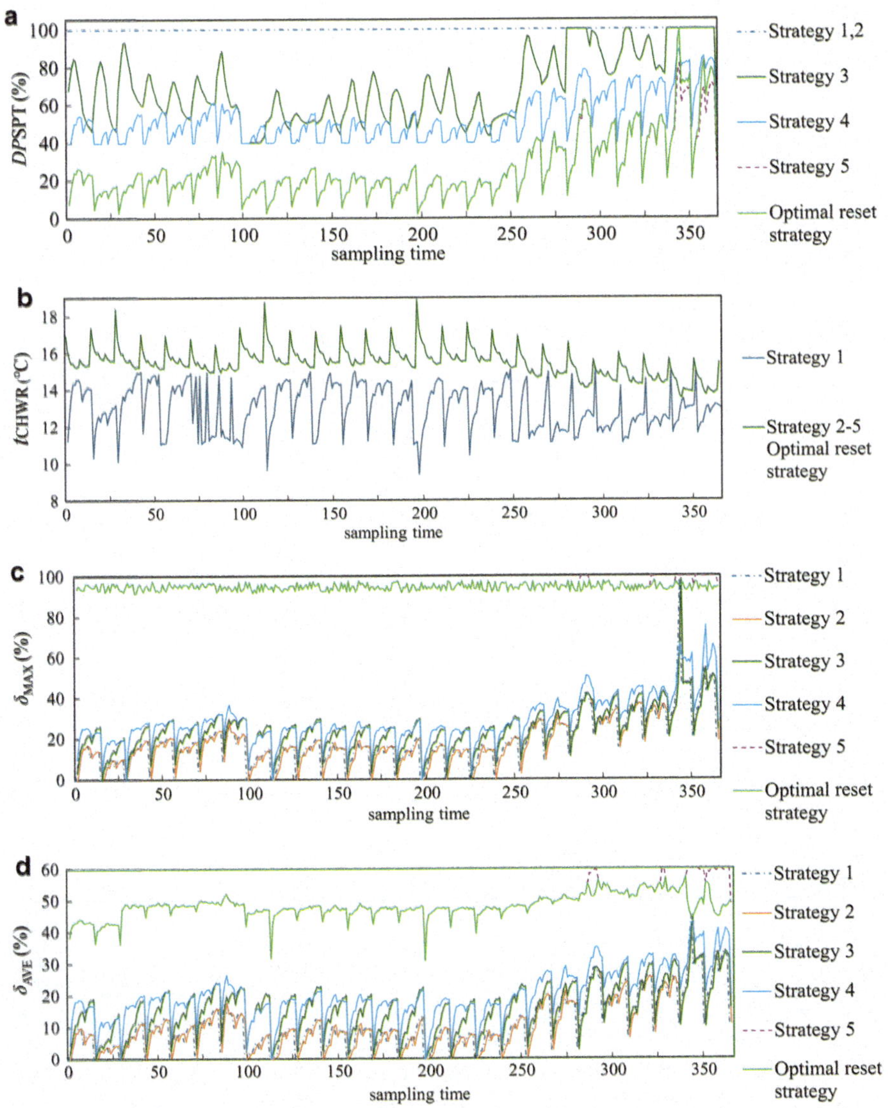

Fig. 5.9 Comparison of variable DP setpoint control test results (redrawn from Ref. [1], copyright 2016, with permission from ELSEVIER)

other hand, the influence of such temperature difference on heat transfer efficiency of evaporator should also be investigated.

Thirdly, the average pump efficiency of all strategies is above 65%, which indicates that the pump optimization control strategy proposed in Sect. 5.2 has a good application effect.

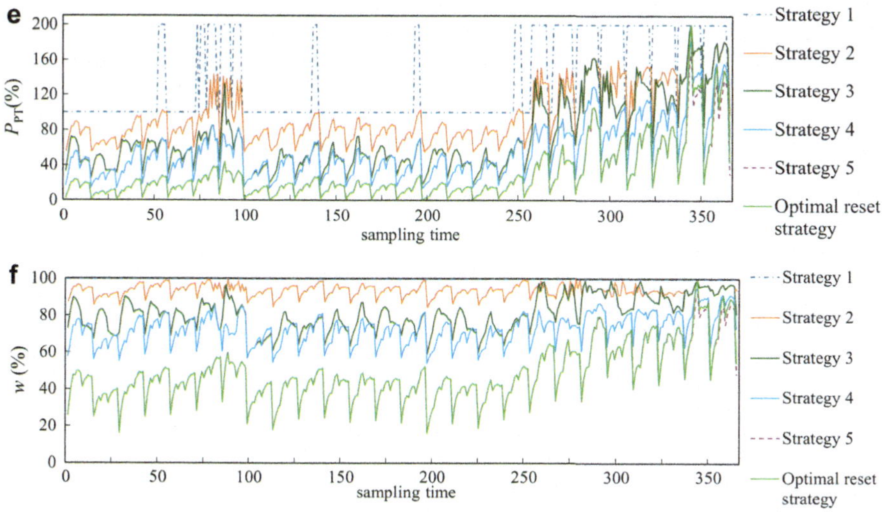

Fig. 5.9 (continued)

Table 5.5 Evaluation parameter test data average values and power consumption (data from ref. [1], copyright 2016, with permission from ELSEVIER)

	DP_{SPT} (%)	t_{CHWR} (°C)	δ_{MAX} (%)	δ_{AVE} (%)	P_{PT} (%)	w (%)	η_P (%)
Strategy 1	100.00	13.00	19.04	11.09	135.99	100.00	70.00
Strategy 2	100.00	15.47	19.04	11.09	97.66	93.97	65.33
Strategy 3	68.99	15.47	24.37	16.07	72.86	81.14	66.70
Strategy 4	53.02	15.47	28.62	19.91	57.77	73.15	66.98
Strategy 5	25.57	15.47	95.28	49.09	30.05	48.25	69.93
Optimal strategy	26.06	15.47	94.85	48.23	30.75	48.46	69.96

Lastly, the application effect of Strategy 5 and optimal reset strategy is very similar, but the δ_{MAX} simulation results of Strategy 5 indicate that, among 364 simulation conditions, δ_{MAX} reached 100% in 30 instances, suggesting the potential occurrence of the MUTL in the system. This issue is evidenced by the valve position distribution frequency depicted in Fig. 5.10. Distribution frequency of valve position is defined as the ratio of the number of hours an AHU operates within a specific valve position range to the total hours during the whole period of simulation.

Several points can be seen from Fig. 5.10:

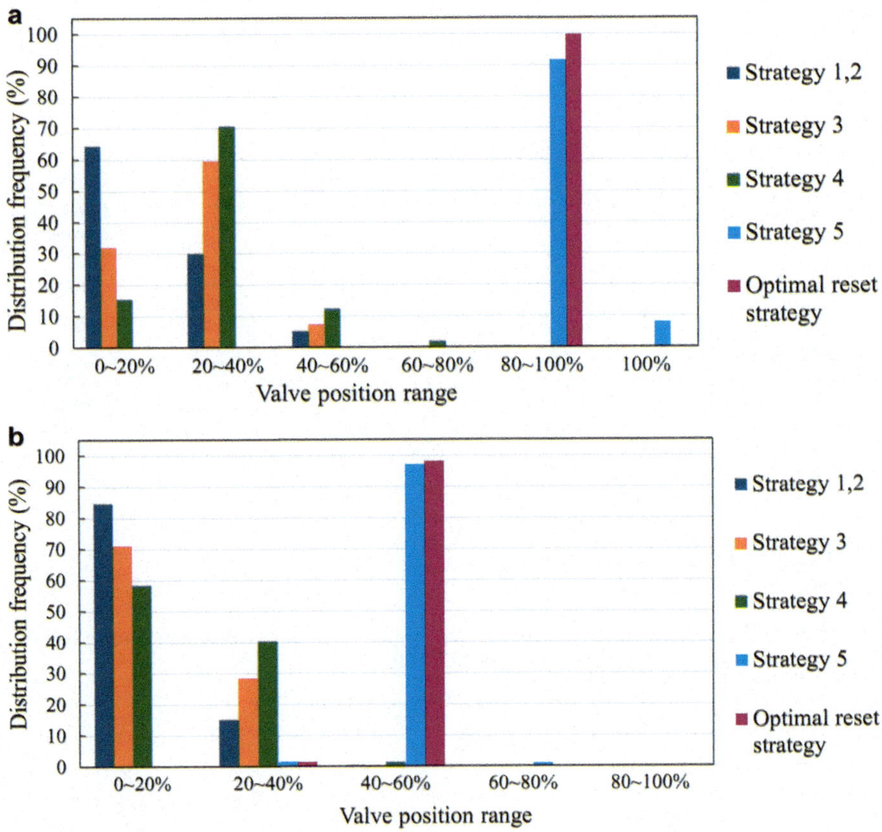

Fig. 5.10 Valve position distribution frequency during test: **a** δ_{MAX} distribution frequency and **b** δ_{AVE} distribution frequency (redrawn from Ref. [1], copyright 2016, with permission from ELSEVIER)

(1) Corresponding to the optimal set valve position domain (80–100%), the distribution frequencies of δ_{MAX} for Strategy 5 and the optimal reset strategy are 91.75% and 100%, respectively, substantially higher than the other four strategies.
(2) Under Strategy 5, δ_{MAX} reaches 100% in 8.24% of the operational conditions. Within these conditions, the maximum valve position loop probably transforms into the MUTL. Contrarily, the optimal reset strategy guarantees δ_{MAX} to be within the optimal valve position domain in all simulation conditions, avoiding the emergence of the MUTLs. This indicates that resetting DP_{SPT} based on the MUHL may not be adaptable to all operational conditions.
(3) The optimal reset strategy ensures that δ_{MAX} stays in a relatively high range in most of the working conditions, reducing the impedance of the pipe network and ensuring each branch's flow supply with a lesser DP_{SPT}.

5.4 Optimal Control for Chilled Water Pumps

The DP_{SPTO} obtained by Eq. (5.3) provides a reasonable DP_{SPT} for the control of chilled water pumps. In the implementation of differential pressure control, for multiple chilled water pumps operating in parallel, there are different operation schemes. The control system is illustrated in Fig. 5.11. The optimal control strategy of water pumps can be described in the following two parts.

First, for the single-pump frequency control, based on the deviation of DP and DP_{SPTO}, the controller uses the appropriate control algorithm to calculate the frequency signal of the variable frequency drive to regulate the pump's operating state. The key point for optimal control lies in the optimization of the control algorithm.

Second, the frequency control of multiple parallel pumps involves the optimization of the number of pump units. In other words, in order to achieve the hydraulic conditions required by current water system (DP_{SPTO} and the required chilled water flow rate), the parallel pump operating mode to achieve the minimal power consumption should be optimized.

In this section, the optimal control strategy for multiple parallel water pumps will be discussed (the second part). Assuming the m parallel water pumps in Fig. 5.11 are of the same model, the following working conditions may exist during the operation.

(1) Both the single-pump operation and multiple-pump parallel operation of can achieve the current system requirements of the hydraulic condition.
(2) During the multiple-pump operation, there are more than one pump configuration schemes corresponding to different speeds can achieve the current system requirements of hydraulic condition.

In the above two conditions, different operating schemes result in varying pump energy consumption, thereby necessitating the optimization of control schemes. The

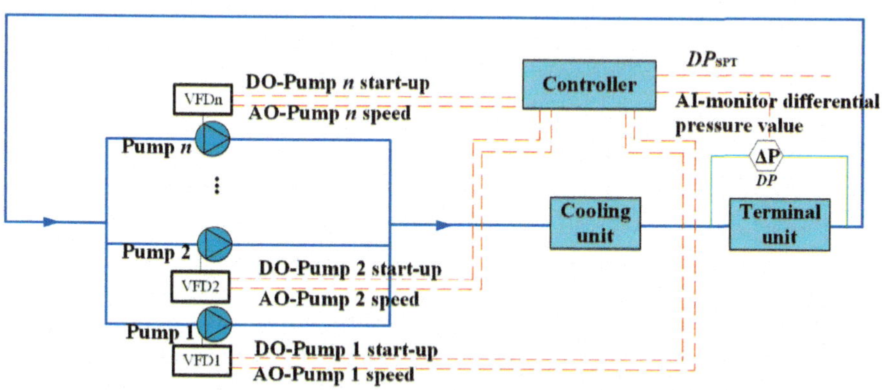

Fig. 5.11 Schematic diagram of a parallel variable frequency pump control system (redrawn from Ref. [3], copyright 2012, with permission from ELSEVIER)

first issue to address is the configuration of the pump speed ratio to minimize energy consumption during multiple-pump operation. The second issue is that corresponding to the required hydraulic condition, whether the single-pump mode or multiple-pump operation mode should be selected to achieve the highest pump efficiency. These two issues are explored in this section.

5.4.1 *The Optimal Allocation of Parallel Variable Frequency Pump Speed Ratio*

It is assumed that m sets of VFD pumps are running in parallel, and the pump operating head required by the system is H_R and the flow rate is Q_R. Suppose the speed ratio of the *i*th pump ($1 < i < m$), flow rate, head, pump efficiency, motor efficiency, and VFD efficiency are w_i, Q_{Pi}, H_{Pi}, η_{Pi}, η_{Mi}, and η_{VFDi}, respectively. The total power of m pumps can be expressed by Eq. (5.13).

$$P_{PT} = \sum_{i=1}^{m} \frac{10 H_{Pi} Q_{Pi}}{\eta_{Pi} \eta_{Mi} \eta_{VFDi}}. \tag{5.13}$$

P_{PT} is the target parameter for the optimal configuration of the speed ratio of the parallel variable frequency pump. According to the operating characteristics of the parallel pump, the following constraints are established.

$$\begin{cases} H_{Pi} H_R \\ Q_R \sum_{i=1}^{m} Q_{Pi} \end{cases}. \tag{5.14}$$

Based on the constraints of Eq. (5.14), P_{PT} can be rewritten as follows:

$$P_{PT} = 10 H_R \sum_{i=1}^{m} \frac{Q_{Pi}}{\eta_{Pi} \eta_{Mi} \eta_{VFDi}}. \tag{5.15}$$

The optimal configuration problem of the speed ratio of the parallel variable frequency pump can be transformed into an optimization problem with the P_{PT} optimization goal and Eq. (5.14) as the constraint condition, i.e., how to configure the speed ratio of each pump to have the least P_{PT} consumption while meeting he same hydraulic operation requirements.

Based on Eq. (5.14), it can be seen that for any pump i in m pumps, if $\frac{Q_{Pi}}{\eta_{Pi}\eta_{Mi}\eta_{VFDi}}$ reaches the minimum, $\sum_{i=1}^{m} \frac{Q_{Pi}}{\eta_{Pi}\eta_{Mi}\eta_{VFDi}}$ also reaches the minimum so that P_{PT} is the minimum. In other words, the optimization objective can be transformed into $\frac{Q_{Pi}}{\eta_{Pi}\eta_{Mi}\eta_{VFDi}}$. Bernier et al. [160] stated that η_M and η_{VFD} only depend on the speed ratio. When $0.4 < w < 1$, η_M and η_{VFD} are less affected by w. The variable range of

$\eta_{\rm M}$ is [92%, 94%], and the variable range of $\eta_{\rm VFD}$ is [83%, 95%]. When $w < 0.4$, w has a significant effect on $\eta_{\rm M}$ and $\eta_{\rm VFD}$. Therefore, if the lower limit of the speed ratio is set to be 0.4 with slight w fluctuation, $\eta_{\rm M}$ and $\eta_{\rm VFD}$ can be considered basically unchanged, and thus, the optimization objective can be simplified as $\frac{Q_{{\rm P}i}}{\eta_{{\rm P}i}}$.

For any two pumps P_i and P_j, combined Eq. (3.13), we find:

$$\begin{cases} \frac{Q_{{\rm P}i}}{\eta_{{\rm P}i}} \frac{1}{b_1 Q_{{\rm P}i} w_{{\rm P}i}^{-2} + b_2 w_{{\rm P}i}^{-1} + b_3 Q_{{\rm P}i}^{-1}} \\ \frac{Q_{{\rm P}j}}{\eta_{{\rm P}j}} \frac{1}{b_1 Q_{{\rm P}j} w_{{\rm P}j}^{-2} + b_2 w_{{\rm P}j}^{-1} + b_3 Q_{{\rm P}j}^{-1}} \end{cases}. \tag{5.16}$$

In Eq. (5.16), the condition that $\frac{Q_{{\rm P}i}}{\eta_{{\rm P}i}}$ and $\frac{Q_{{\rm P}j}}{\eta_{{\rm P}j}}$ reach the local minimum simultaneously is that the denominator on the right side of the formula reaches the local maximum value. Taking Pi as an example, that is, $b_1 Q_{{\rm P}i} w_{{\rm P}i}^{-2} + b_2 w_{{\rm P}i}^{-1} + b_3 Q_{{\rm P}i}^{-1}$ reaches the local maximum value, $W_i = w_{{\rm P}i}^{-1}$. The W_i that meets the extreme value condition can be expressed as $-\frac{b_2}{2b_1 Q_{{\rm P}i}}$, then there is

$$w_{{\rm P}i} = -\frac{2b_1 Q_{{\rm P}i}}{b_2}. \tag{5.17}$$

Substitute $Q'_{\rm PR} = -\frac{b_2}{2b_1}$ (the pump flow corresponding to the maximum pump efficiency, as expressed in Eq. (3.9)) into Eq. (5.17), there is:

$$w_{{\rm P}i} = \frac{Q_{{\rm P}i}}{Q'_{PR}}. \tag{5.18}$$

This means that when Eq. (5.18) holds, $\frac{Q_{{\rm P}i}}{\eta_{{\rm P}i}}$ will reach the local minimum value. Referring to the change range of $w_{{\rm P}i}$, there is:

(1) When $Q_{{\rm P}i} \le Q'_{\rm PR}$, $w_{{\rm P}i} \le 1$. It fits the actual range of speed ratio, and the w_{Pi} corresponding to the local minimum value of $\frac{Q_{{\rm P}i}}{\eta_{{\rm P}i}}$ is $-\frac{2b_1 Q_{{\rm P}i}}{b_2}$. Similarly, when $Q_{Pj} Q'_{\rm PR}$, the $w_{{\rm P}j}$ corresponding to the local minimum value of $\frac{Q_{{\rm P}j}}{\eta_{{\rm P}j}}$ is $-\frac{2b_1 Q_{{\rm P}j}}{b_2}$. Combining Eq. (3.12) with Eq. (5.14), there is:

$$a_1 Q_{{\rm P}i}^2 + a_2 w_{{\rm P}i} Q_{{\rm P}i} + a_3 w_{{\rm P}i}^2 = a_1 Q_{{\rm P}j}^2 + a_2 w_{{\rm P}j} Q_{{\rm P}j} + a_3 w_{Pj}^2. \tag{5.19}$$

Substitute $w_{{\rm P}i} = -\frac{2b_1 Q_{{\rm P}i}}{b_2}$ and $w_{{\rm P}j} = -\frac{2b_1 Q_{{\rm P}j}}{b_2}$ into Eq. (5.19), it can be found that

$$\left(a_1 - \frac{2a_2 b_1}{b_2} + \frac{4a_3 b_1^2}{b_2^2}\right)(Q_{{\rm P}i} + Q_{{\rm P}j})(Q_{{\rm P}i} - Q_{{\rm P}j}) = 0. \tag{5.20}$$

Set $C_1 = a_1 - \frac{2a_2 b_1}{b_2} + \frac{4a_3 b_1^2}{b_2^2}$. Based on the size relationship of a_1, a_2, a_3, b_1, and b_2, we can know C_1 is not 0. Therefore, the established condition that $P_{\rm PT}$ reaches local minimum value is that Eq. (5.20) holds, i.e., $w_{{\rm P}i} = w_{{\rm P}j}$.

(2) When $Q_{Pi} > Q'_{PR}$, $w_{Pi} > 1$. It is beyond the actual range of the speed ratio, the w_{Pi} corresponding to the local minimum of $\frac{Q_{Pi}}{\eta_{Pi}}$ is taken to be 1. Similarly, when $Q_{Pj} > Q'_{PR}$, the w_{Pj} corresponding to the minimum of $\frac{Q_{Pj}}{\eta_{Pj}}$ is taken to be 1.

Substitute $w_{Pi} = 1$, $w_{Pj} = 1$ into Eq. (5.19), we can find:

$$\left(Q_{Pi} - Q_{Pj}\right)\left[a_1\left(Q_{Pi} + Q_{Pj}\right) + a_2\right] = 0. \tag{5.21}$$

Assuming that $C_2 = a_1\left(Q_{Pi} + Q_{Pj}\right) + a_2$, based on the size relationship of a_1 and a_2, we can know that C_2 is not 0. Therefore, the establishment requirement for Eq. (5.21) is $Q_{Pi} = Q_{Pj}$. According to Eqs. (3.12) and (5.14), we can know that $w_{Pi} = w_{Pj}$. The condition that P_{PT} reaches the local minimum is that Eq. (5.21) holds, i.e., $w_{Pi} = w_{Pj}$.

(3) When $Q_{Pi} \le Q'_{PR}$, $w_{Pi} \le 1$. It is within the actual range of the speed ratio, and the w_{Pi} corresponding to the minimum value of $\frac{Q_{Pi}}{\eta_{Pi}}$ is $-\frac{2b_1 Q_{Pi}}{b_2}$. When $G_{Pj} > G'_{PR}$, w_{Pj} corresponding to the minimum value of $\frac{Q_{Pj}}{\eta_{Pj}}$ is 1.

Substitute $w_{Pi} = -\frac{2b_1 Q_{Pi}}{b_2}$, $w_{Pj} = 1$, $Q'_{PR} = -\frac{b_2}{2b_1}$ into Eq. (5.19), it can be found that

$$a_1\left(Q_{Pi} + Q_{Pj}\right)\left(Q_{Pi} - Q_{Pj}\right) + a_2\left(\frac{Q_{Pi}^2}{Q'_{PR}} - Q_{Pj}\right) + a_3\left(\frac{Q_{Pi}^2}{Q'_{PR}} - 1\right) = 0. \tag{5.22}$$

Based on the size relationship of a_1, a_2 and a_3, the established condition for Eq. (5.22) is $Q_{Pi} = Q_{Pj} = Q'_{PR}$. Since $Q_{Pj} > Q'_{PR}$, $\frac{Q_{Pi}}{\eta_{Pi}}$ and $\frac{Q_{Pj}}{\eta_{Pj}}$ cannot reach the local minimum at the same time under current working condition.

(4) When $Q_{Pi} > Q'_{PR}$, $w_{Pi} > 1$. It exceeds the actual range of speed ratio, and the w_{Pi} corresponding to the local minimum value of $\frac{Q_{Pi}}{\eta_{Pi}}$ is 1. When $Q_{Pj} \le Q'_{PR}$, w_{Pj} corresponding to the local minimum value of $\frac{Q_{Pj}}{\eta_{Pj}}$ is $-\frac{2b_1 Q_{Pj}}{b_2}$.

Substitute $w_{Pi} = 1$, $w_{Pj} = -\frac{2b_1 Q_{Pj}}{b_2}$ and $Q'_{PR} = -\frac{b_2}{2b_1}$ into Eq. (5.19), it can be found that

$$a_1\left(Q_{Pi} + Q_{Pj}\right)\left(Q_{Pi} - Q_{Pj}\right) + a_2\left(\frac{Q_{Pj}^2}{Q'_{PR}} - Q_{Pi}\right) + a_3\left(\frac{Q_{Pj}^2}{Q'^2_{PR}} - 1\right) = 0. \tag{5.23}$$

Based on the size relationship of a_1, a_2, and a_3, the establishment condition of Eq. (5.23) is $Q_{Pi} = Q_{Pj} = Q'_{PR}$. Since $Q_{Pi} > Q'_{PR}$, $\frac{Q_{Pi}}{\eta_{Pi}}$ and $\frac{Q_{Pj}}{\eta_{Pj}}$ cannot reach the local minimum value at the same time under current working condition.

In summary, it can be known from the above analysis that for any two pumps in the parallel variable frequency pump group, corresponding the same hydraulic working condition in the same system, the condition to achieve the lowest power is having the same speed ratio.

5.4.2 The Optimal Unit Number for a Parallel Variable Frequency Water Pump

Assuming there are n sets of parallel variable frequency pumps operating in the system, the required head and flow rate is H_{R} and Q_{R}, respectively. Both following two configuration options can achieve the requirement.

Option 1: One single pump operates at speed of w_1 and the power is P_{P1}.

$$P_{P1} = \frac{10 H_{\mathrm{R}} Q_{\mathrm{R}}}{\eta_1 \eta_{\mathrm{M1}} \eta_{\mathrm{VFD1}}}. \tag{5.24}$$

In the equation, η_1 denotes the pump efficiency, $\eta_1 = b_1 w_1^{-2} Q_{\mathrm{R}}^2 + b_2 w_1^{-1} Q_{\mathrm{R}} + b_3$. η_{M1} is the motor efficiency, and the η_{VFD} is the efficiency of the variable frequency drive.

Option 2: m pumps($1 < m \leq n$) operate at the speed of w_{m}, and the total power of pumps is $P_{\mathrm{P}m}$.

$$P_{\mathrm{P}m} = \sum_{i=1}^{m} \frac{10 H_{\mathrm{R}} Q_{\mathrm{P}i}}{\eta_m \eta_{\mathrm{M}_m} \eta_{\mathrm{VFD}_m}} = \frac{10 H_{\mathrm{R}} Q_{\mathrm{R}}}{\eta_m \eta_{\mathrm{M}_m} \eta_{\mathrm{VFD}_m}}. \tag{5.25}$$

In the equation, $Q_{\mathrm{P}i}$ represents the flow rate for one single pump with option 2, $Q_{\mathrm{P}i} = \frac{Q_{\mathrm{R}}}{m}$. η_m represents the water pump efficiency with option 2, $\eta_m = b_1 w_m^{-2} Q_{\mathrm{P}i}^2 + b_2 w_m^{-1} Q_{\mathrm{P}i} + b_3$. $\eta_{\mathrm{M}m}$ is the motor efficiency, and the $\eta_{\mathrm{VFD}m}$ is the efficiency of the variable frequency drive. The range of w_m will be discussed in the following part.

According to Eq. (3.12) and Eq. (5.14), we can find:

$$a_1 Q_{\mathrm{R}}^2 + a_2 w_1 Q_{\mathrm{R}} + a_3 w_1^2 = a_1 \left(\frac{Q_{\mathrm{R}}}{m}\right)^2 + a_2 w_m \left(\frac{Q_{\mathrm{R}}}{m}\right) + a_3 w_m^2. \tag{5.26}$$

Since $m > 1$, according to Eq. (5.26), there is $w_m < w_1$. Based on Eq. (5.14), there is

$$H_{\mathrm{R}} < a_3 w_m^2. \tag{5.27}$$

i.e., the range of w_m is

$$\sqrt{\frac{H_{\mathrm{R}}}{a_3}} < w_m < w_1. \tag{5.28}$$

Equations (5.14) and (5.28) are the constraints for the content discussed in this section. The following is the optimization of the above two configuration schemes by comparing the size of P_{P1} and P_{Pm}.

Since most of the characteristic curves of the pumps used in the HVAC field are relatively flat, w_1 is not very different from w_m. As mentioned above, the impact of η_M and η_{VFD} on P_P can be ignored when the speed ratio is higher than 0.4 and fluctuates slightly. Therefore, the value of N_{P1} and N_{Pm} mainly depends on the value of η_1 and η_m. Set $\Delta\eta = \eta_1 - \eta_m$. Combined with the expression η_1 and η_m, there is

$$\Delta\eta = -b_1 Q_R^2 M^2 - b_2 Q_R M + b_1 \left(\frac{Q_R}{w_1}\right)^2 + b_2 \frac{Q_R}{w_1}. \tag{5.29}$$

In the equation, $M = \frac{1}{m w_m}$, based on Eq. (5.28), there is

$$\frac{1}{m w_1} < M < \frac{1}{m}\sqrt{\frac{a_3}{H_R}}. \tag{5.30}$$

Corresponding to Eq. (5.29), when $M = -\frac{b_2}{2b_1 Q_R}$, the local minimum of $\Delta\eta$ can be obtained as $\Delta\eta_{MIN}$.

$$\Delta\eta_{MIN} = b_1 w_1^{-2} Q_R^2 + b_2 w_1^{-1} Q_R + \frac{b_2^2}{4b_1}. \tag{5.31}$$

Combined with the expression equation of η_1, there is

$$\Delta\eta_{MIN} = \eta_1 - b_3 + \frac{b_2^2}{4b_1}. \tag{5.32}$$

(1) When $\eta_1 > b_3 - \frac{b_2^2}{4b_1}$,

$\Delta\eta_{MIN} > 0$, i.e., $\eta_1 > \eta_m$. Therefore, option 1 is more advantageous in terms of energy conservation. However, through the expression of η_1, it can be found that $b_3 - \frac{b_2^2}{4b_1}$ is the local maximum value corresponding to η_1. Therefore, $\eta_1 > b_3 - \frac{b_2^2}{4b_1}$ is not valid.

(2) When $\eta_1 \le b_3 - \frac{b_2^2}{4b_1}$,

$\Delta\eta_{MIN} \leqq 0$, and it is hard to determine the size relationship of η_1 and η_m. Assuming the M value corresponding to $\Delta\eta_{MIN}$ is M_O and the range of M is $[M_{MIN}, M_{MAX}]$, there is

$$\begin{cases} M_O \frac{b_2}{2b_1 Q_R} \\ M_{MIN} \frac{1}{m w_1} \\ M_{MAX} \frac{1}{m}\sqrt{\frac{a_3}{H_R}} \end{cases}. \tag{5.33}$$

(1) When $M_O < M_{MIN}$, i.e., $Q_R > -\frac{b_2 m w_1}{2b_1}$, the range of M is in the monotonically increasing region to the right of the extreme point. Combined with the expression

for η_1, the local maximum and minimum values of $\Delta\eta$ can be expressed as Eqs. (5.34) and (5.35), respectively.

$$\Delta\eta_{\mathrm{MIN}} = \left(1 - m^{-1}\right)\left(\eta_1 - b_3 + b_1 m^{-1} w_1^{-2} Q_{\mathrm{R}}^2\right). \tag{5.34}$$

$$\Delta\eta_{\mathrm{MAX}} = \eta_1 - b_3 - b_1 Q_{\mathrm{R}}^2 \left(m\sqrt{\frac{H_{\mathrm{R}}}{a_3}}\right)^{-2} - b_2 Q_{\mathrm{R}} \left(m\sqrt{\frac{H_{\mathrm{R}}}{a_3}}\right)^{-1}. \tag{5.35}$$

When $\Delta\eta_{\mathrm{MIN}} > 0$, i.e., $\eta_1 > b_3 - b_1 w_1^{-2} Q_{\mathrm{R}}^2 m^{-1}$, $\eta_1 > \eta_m$, which means that the option 1 is more energy-efficient.

When $\Delta\eta_{\mathrm{MAX}} < 0$, i.e., $\eta_1 < b_3 + b_1 Q_{\mathrm{R}}^2 (m\sqrt{\frac{H_{\mathrm{R}}}{a_3}})^{-2} + b_2 Q_{\mathrm{R}} \left(m\sqrt{\frac{H_{\mathrm{R}}}{a_3}}\right)^{-1}$, there is $\eta_1 < \eta_m$, which means that the option 2 is more energy-efficient.

(2) When $M_{\mathrm{O}} \in [M_{\mathrm{MIN}}, M_{\mathrm{MAX}}]$, i.e., $Q_{\mathrm{R}} \in \left[-\frac{b_2 m}{2b_1}\sqrt{\frac{H_{\mathrm{R}}}{a_3}}, -\frac{b_2 w_1 m}{2b_1}\right]$, we compare the $\Delta\eta_{\mathrm{MIN}}$ and $\Delta\eta_{\mathrm{MAX}}$ in Eqs. (5.34) and (5.35), respectively, assuming the maximum of the two is $\Delta\eta_{\mathrm{M}}$. When $\Delta\eta_{\mathrm{M}} < 0$, there is $\eta_1 < \eta_m$, which means that the option 2 is more energy-efficient.

(3) When $M_{\mathrm{O}} > M_{\mathrm{MAX}}$, i.e., $Q_{\mathrm{R}} < -\frac{b_2 m}{2b_1}\sqrt{\frac{H_{\mathrm{R}}}{a_3}}$, the range of M is in the monotone decreasing region to the right of the extreme point. At this time, $\Delta\eta_{\mathrm{MIN}}$ and $\Delta\eta_{\mathrm{MAX}}$ in Eqs. (5.34) and (5.35) are, respectively, the local maximum and local minimum values of $\Delta\eta$.

When $\Delta\eta_{\mathrm{MAX}} > 0$, i.e., $\eta_1 > b_3 + b_1 Q_{\mathrm{R}}^2 (m\sqrt{\frac{H_{\mathrm{R}}}{a_3}})^{-2} + b_2 Q_{\mathrm{R}} \left(m\sqrt{\frac{H_{\mathrm{R}}}{a_3}}\right)^{-1}$, $\eta_1 > \eta_m$, which means that the option 1 is more energy-efficient.

When $\Delta\eta_{\mathrm{MIN}} < 0$, i.e., $\eta_1 < b_3 - b_1 w_1^{-2} Q_{\mathrm{R}}^2 m^{-1}$, there is $\eta_1 > \eta_m$, which means that the option 2 is more energy-efficient.

(4) When the current working condition fails to satisfy the above requirements, considering the fault possibility of pump under multiple pumps working mode is relatively high, the option 1 is the preferred option.

In summary, the two configuration options can be optimized by referring to the range of system flow Q_{R} and pump efficiency η_1 corresponding to configuration option 1. The above derivation process is shown in the pump performance curve in the form of working condition partitions, as shown in Fig. 5.12.

In Fig. 5.12, $Q_{\mathrm{MIN}} = -\frac{b_2 m}{2b_1}\sqrt{\frac{H_{\mathrm{R}}}{a_3}}$, $Q_{\mathrm{MAX}} = -\frac{b_2 m w_1}{2b_1}$,

$\eta_A = b_3 - b_1 w_1^{-2} Q_{\mathrm{R}}^2 m^{-1}$, $\eta_B = b_3 + b_1 Q_{\mathrm{R}}^2 \left(m\sqrt{\frac{H_{\mathrm{R}}}{a_3}}\right)^{-2} + b_2 Q_{\mathrm{R}} \left(m\sqrt{\frac{H_{\mathrm{R}}}{a_3}}\right)^{-1}$.

Q_{MIN} and Q_{MAX} divide the system flow into three zones: $Q1$, $Q2$, and $Q3$. η_A and η_B divide the pump efficiency into three zones: $\eta 1$, $\eta 2$, and $\eta 3$. The zone boundaries are shown by bold dashed lines in Fig. 5.12. Table 5.6 lists the working conditions for each zone and the corresponding optimal configuration scheme.

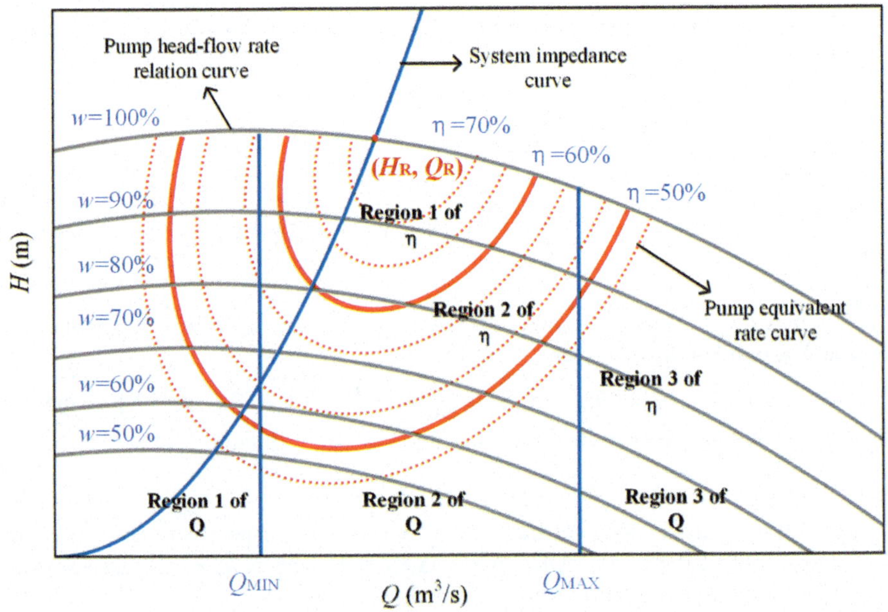

Fig. 5.12 Reference condition of optimal pump control strategy (redrawn from Ref. [3], copyright 2012, with permission from ELSEVIER)

Table 5.6 Condition partition of optimal allocation of operation number in parallel variable speed pump system (data from Ref. [3], copyright 2012, with permission from ELSEVIER)

Flow region	Working condition	Efficiency region	Working condition	Optimal configuration
Region 1 of Q	$Q_R < Q_{MIN}$	Region 1 of η	$\eta_1 > \eta_B$	Option 1
		Region 2 of η	$\eta_1 \in [\eta_A, \eta_B]$	Option 1
		Region 3 of η	$\eta_1 < \eta_A$	Option 2
Region 2 of Q	$Q_R \in [Q_{MIN}, Q_{MAX}]$	Region 1 of η, Region 2 of η	$\eta_1 \geqq MIN(\eta_A, \eta_B)$	Option 1
		Region 3 of η	$\eta_1 < MIN(\eta_A, \eta_B)$	Option 2
Region 3 of Q	$Q_R > Q_{MAX}$	Region 1 of η	$\eta_1 > \eta_A$	Option 1
		Region 2 of η	$\eta_1 \in [\eta_B, \eta_A]$	Option 1
		Region 3 of η	$\eta_1 < \eta_B$	Option 2

Since Q_{MIN}, Q_{MAX}, η_A, and η_B are closely related to the hydraulic condition of the system, the zoning conditions in the figure vary with the working conditions. As can be seen in the figure, the current working point of the pump (H_R, Q_R) is in the region 2 of Q and region 1 of η. According to Table 5.6, the optimal scheme under current working condition is operating with one single pump.

In summary, the optimal configuration method proposed in this section provides a simple and effective reference for deciding the unit number of parallel variable frequency pumps. The implementation of the method only needs to satisfy the following two conditions.

First, an accurate pump characteristic model is required, i.e., a_1, a_2, a_3, b_1, b_2, and b_3 in Eqs. (3.8) and (3.9) are determined.

Second, the pump head and system flow rate can be continuously monitored, denoted by H_R and Q_R.

As for w_1 in the calculation formula of η_A, if option 1 is adopted, w_1 can be obtained directly through the pump frequency feedback signal. If option 2 is adopted, w_1 can be obtained by Eq. (3.12).

5.5 Summary

In this chapter, based on the definition and identification method of the MUTL, the optimal control strategy of the DP setpoint is proposed based on the maximum valve position loop valve position. Secondly, aiming at the stability problem caused by the water system regulation process, the energy-saving effect, hydraulic stability, and thermodynamic stability indexes of several pressure difference control strategies in a specific regulation period are studied by simulation. Thirdly, the application of six DP_{SPT} resetting strategies, including optimal resetting strategy, is studied by simulation, and the characteristics of each strategy are analyzed from the aspects of energy-saving effect and satisfying level of users' demand. Finally, the optimal control strategy of water pumps is studied, and the optimal configuration method of speed and number of parallel variable frequency water pump is proposed.

(1) The maximum valve position loop is an important reference loop to determine the possible position of the MUTL in advance. The optimal valve position domain and the optimal valve position domain provide an effective quantitative reference for the research on the optimal reset of DP setpoint.
(2) The active control of the pump can improve the thermodynamic stability of the pipe network, and the appropriate selection of the DP setpoint adjustment step can improve and balance the thermodynamic stability of each branch.
(3) Compared with other DP_{SPT} resetting strategies, the optimal resetting strategy proposed in this chapter can obtain ideal pump energy-saving effect and avoid the appearance of the MUHL on the other hand.
(4) For the same pump operating point required by the DP setpoint, the optimal control strategy of parallel variable frequency pump proposed in Sect. 5.4 can significantly improve the pump operating efficiency and reduce the pump operating power consumption.

References

1. Tianyi Z, Liangdong M, Jili Z (2016) An optimal differential pressure reset strategy based on the most unfavorable thermodynamic loop on-line identification for a variable water flow air conditioning system. Energy Build 110:257–268. https://doi.org/10.1016/j.enbuild.2015.10.032
2. Tianyi Z, Jili Z, Liangdong M (2013) Experimental analysis of the thermal stability of the pressure control method for a variable flow air-conditioning water system. Build Environ 70:1–9. https://doi.org/10.1016/j.buildenv.2013.07.012
3. Zhao T, Zhang J, Ma L (2012) On-line optimization control method based on extreme value analysis for parallel variable-frequency hydraulic pumps in central air-conditioning systems. Build Environ 47:330–338. https://doi.org/10.1016/j.buildenv.2011.07.007

Chapter 6
Hysteresis and Nonlinear Characteristics of the System

Besides optimization of pipe network, the HVAC water systems also exhibit significant hysteresis and nonlinear characteristics when undergoing optimized control, which is of important reference significance for improving the decoupling optimization effect of the system. The hysteresis feature refers to the time-delay response characteristics exhibited by the parameters when propagating in the pipeline network, while the nonlinear feature refers to the spatial coupling characteristics of interactions between devices. This chapter will conduct a model study on the hysteresis and nonlinear characteristics of HVAC water systems, constructing the time-delay response characteristics of parameters in the pipeline network and the nonlinear characteristics among various devices. The chapter will first analyze the dynamic response characteristics of water temperature and flow in the HVAC water system pipeline network, exploring the time-delay characteristics of the quality–quantity parameters in the control process. Secondly, it will analyze the influence scale of constitutive parameters on the parameter time-delay response characteristics, verifying the applicability of the time-delay characteristics model in systems of different volumes and structures. Finally, by considering the coupling characteristics among operating parameters of various devices in the system, the chapter will conduct a model study on the spatial characteristics of the system, clarifying the spatial coupling relationships among various devices and control loops. The characteristic model of the HVAC water system established in this chapter will lay the theoretical foundation for subsequent chapters' research on global optimization methods.

6.1 Hysteresis Characteristic of System Parameters

The specific manifestation of parameter hysteresis characteristics in the pipe network can be divided into thermal and hydraulic characteristics according to the type of parameter. For the thermal dynamic response characteristics, the energy balance

T. Zhao et al., *Thermal Distribution Principle and Nonlinear Optimal Control for HVAC Water System*, Indoor Environment and Sustainable Building,
https://doi.org/10.1007/978-981-96-0377-0_6

equation can be established for the fluid element using the finite volume method, and the pipeline thermal response characteristics can be solved. Then, by combining graph theory and the energy balance equation of pipe network, the thermal dynamic response characteristics of the entire pipeline network can be obtained. For the hydraulic dynamic response characteristics, the fluid inside the pipe can be modeled based on the fluid motion differential equation and energy conservation equation. Similarly, the non-steady-state hydraulic characteristics modeling of the pipeline network can be completed by combining graph theory. The details are illustrated as follows.

6.1.1 Hysteresis Characteristics of Temperature

For the temperature field of the incompressible fluid inside the pipe, the energy balance analysis can be carried out based on a fluid element, as shown in Fig. 6.1. According to the energy balance of the fluid element, the following first-order partial differential equation [1, 2] can be established:

$$\rho c_p A \frac{\partial T}{\partial t} + m c_p \frac{\partial T}{\partial x} = k_w A \frac{\partial^2 T}{\partial x^2} + \frac{T_s - T}{R}. \tag{6.1}$$

where A is the cross-sectional area; m is the fluid mass flow rate; k_w is the fluid thermal conductivity; T_s is the environmental temperature; R is the comprehensive heat transfer resistance, including the convective heat transfer resistance between the fluid and the pipe wall, the conductive heat transfer resistance from the pipe wall to the outer wall, the conductive heat transfer resistance from the outer wall to the insulation layer, and the convective heat transfer resistance from the insulation layer to the environment.

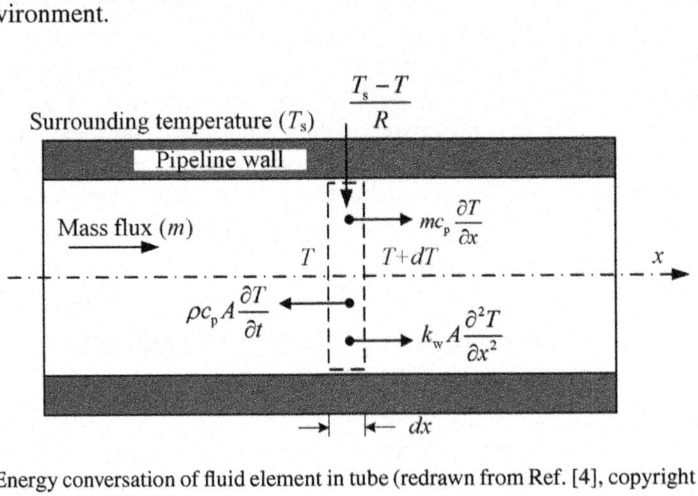

Fig. 6.1 Energy conversation of fluid element in tube (redrawn from Ref. [4], copyright 2022, with permission from ELSEVIER)

The first to fourth terms of the above equation represent the internal energy of the fluid, the enthalpy flux of the fluid, the internal conductive heat transfer of the fluid, and the convective heat transfer of the fluid relative to the outside of the pipe wall, respectively. Since the internal conductive heat transfer term of the fluid can be ignored relative to other terms [3], the above equation can be simplified by omitting the third term:

$$\frac{\partial T}{\partial t} + \frac{m}{\rho A}\frac{\partial T}{\partial x} = \frac{T_s - T}{\rho c_p A R}. \tag{6.2}$$

The above equation is the temperature response characteristic model inside the pipeline. When the environmental temperature (T_s), the initial temperature distribution along the pipeline, and the inlet temperature variation of the pipeline are given, the hourly temperature distribution of the fluid along the pipeline can be obtained by solving the equation with the numerical solution method.

Herein, the characteristic line method, which is commonly used to solve first-order partial differential equations, is used to solve the model. Thus, the model can be represented as the following discrete equation:

$$T_i^{n+1} = \omega \Delta t T_s + (1 - \omega \Delta t)\left(aT_{i-2}^n + bT_{i-1}^n + cT_i^n + dT_{i+1}^n\right). \tag{6.3}$$

where, the subscript n represents the time series, and i represents the spatial sequence; the coefficients a–d are as follows:

$$a = \frac{(\Delta x - v^n \Delta t)(-v^n \Delta t)(-\Delta x - v^n \Delta t)}{-6\Delta x^3}, \tag{6.4}$$

$$b = \frac{(2\Delta x - v^n \Delta t)(-v^n \Delta t)(-\Delta x - v^n \Delta t)}{2\Delta x^3}, \tag{6.5}$$

$$c = \frac{(2\Delta x - v^n \Delta t)(\Delta x - v^n \Delta t)(-\Delta x - v^n \Delta t)}{-2\Delta x^3}, \tag{6.6}$$

$$d = \frac{(2\Delta x - v^n \Delta t)(\Delta x - v^n \Delta t)(-v^n \Delta t)}{6\Delta x^3}. \tag{6.7}$$

where Δt and Δx are the time step and space step, respectively, and ω can be represented as:

$$\omega = \frac{1}{\rho c_p A R}. \tag{6.8}$$

Therefore, based on the flow velocity inside the pipe and the temperature distribution along the pipe at the previous moment, the temperature distribution inside the pipe at the current moment can be solved. To ensure the stability of the numerical solution process, the time step and space step of the characteristic line method should satisfy the Courant–Friedrichs–Lewy (CFL) condition [5, 6], that is:

$$\frac{v^n \Delta t}{\Delta x} < 1. \tag{6.9}$$

The physical meaning of this condition is that the speed at which the solution is advanced in time ($\Delta x/\Delta t$) must be smaller than the fluid disturbance propagation speed v, which ensures that the solving process can capture all the propagating disturbances.

6.1.2 Hysteresis Characteristics of Water Flow

When optimizing and controlling the HVAC water system, the state of the water flow in the pipeline network will gradually transition from the current steady-state condition to the next steady-state condition. During this period, the flow rate of the fluid and the pressure distribution in the pipeline will change over time. The non-steady-state change process of the water flow can be described by the fluid motion equation and the energy equation, which can be used to explore the time-delay characteristics in the propagation process.

The force analysis of non-steady-state fluid control volume is shown in Fig. 6.2. In which, the fluid flows in the x-direction, α is the angle between the pipeline and the horizontal reference line, A is the cross-sectional area at position x, and $A_x\delta_x$ is the elemental volume. Assuming that the fluid flow inside the pipe is homogeneous one-dimensional flow, that is, the density, flow velocity, and pressure on any cross-section are uniformly distributed. Along the direction of fluid flow, according to Newton's second law of motion, the following equations can be derived:

$$\begin{aligned} pA - \left[pA + (pA)_x\delta_x\right] + \left(p + p_x\frac{\delta_x}{2}\right)A_x\delta_x \\ - \tau_0\pi D\delta_x - \rho g A\delta_x \sin\alpha = \rho A\delta_x\dot{v} \end{aligned} \tag{6.10}$$

where $p(x, t)$ is the average pressure of the cross-section, $v(x, t)$ is the average flow velocity of the cross-section; the left-hand side of the equation represents the total pressure on the cross-section, the pressure acting on the annular surface, the shear force acting on the wall surface, and the gravitational force of the fluid control volume in the x-direction. The right-hand side represents the product of the mass and acceleration of the control volume. After omitting higher-order terms containing $(\delta_x)^2$ and expanding the $(pA)_x$ term, the equation can be simplified to:

$$p_x A + \tau_0\pi D + \rho g A \sin\alpha + \rho A\dot{v} = 0. \tag{6.11}$$

Combining the static equilibrium equation, it can be obtained [7]:

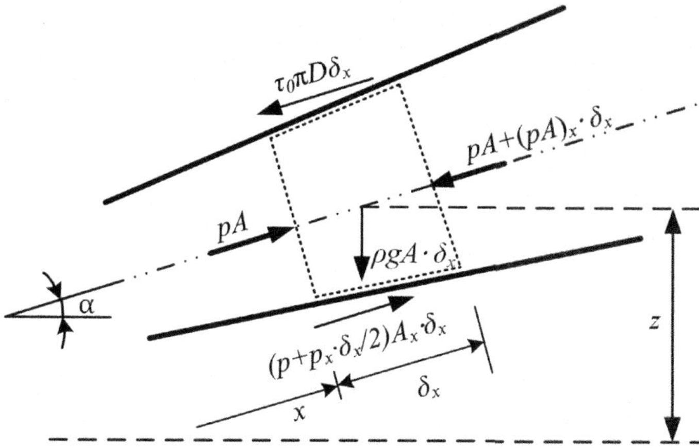

Fig. 6.2 Schematic diagram of fluid element force analysis (redrawn from Ref. [4], copyright 2022, with permission from ELSEVIER)

$$\frac{p_x}{\rho} + v \cdot v_x + v_t + g \cdot \sin\alpha + \frac{fv|v|}{2D} = 0. \tag{6.12}$$

For an incompressible fluid, v_x can be approximated as 0; when the pipe is placed horizontally, i.e., $\alpha = 0$, integrating both sides of the equation from $x = 0$ to L, the following can be obtained:

$$v_t = -f\frac{v|v|}{2D} + \frac{\Delta p}{\rho L}, \quad \Delta p = p_0 - p_L \tag{6.13}$$

where p_0 is the inlet pressure and p_L is the outlet pressure. Assuming that there is no change in the flow direction during the unsteady process, multiplying the flow velocity in the equation by the cross-sectional area, the hydraulic model of the unsteady-state condition of the horizontal pipeline can be obtained:

$$Q_t = -\frac{2f}{\pi D^3}Q^2 + \frac{\pi D^2}{4\rho L}\Delta p. \tag{6.14}$$

By combining the graph theory, the unsteady hydraulic characteristics model of the entire pipeline network can be obtained. The flow balance equation and the pressure balance equation of the pipeline network can be obtained:

$$B_r \cdot \tilde{Q}(t) = 0, \tag{6.15}$$

$$C_r \cdot \Delta\tilde{P}(t) = 0. \tag{6.16}$$

where $\tilde{Q}(t)$ represents the flow collection at time t, and $\Delta\tilde{P}(t)$ represents the pressure difference collection on each side at time t. According to the hydraulic model of the pipeline in the unsteady-state equation (6.14), combined with the hydraulic characteristics of valves, water pumps, and other equipment, the unsteady-state hydraulic model of pipeline i can be described as:

$$K_{f,i} \cdot \dot{Q}_i = -[S_i + \phi_i(x_i)]|\tilde{Q}_i|\tilde{Q}_i + H_{p,i}(n_i, \tilde{Q}_i) + \Delta\tilde{P}_i. \tag{6.17}$$

where $K_{f,i}$ is denoted as:

$$K_{f,i} = \frac{4\rho L_i}{\pi D_i^2}. \tag{6.18}$$

$\dot{Q}_i$ is the derivative of the flow rate of pipeline i with respect to time t; $\tilde{Q}_i$ is the flow rate of pipeline i; $\Delta\tilde{P}_i$ is the pressure difference between the two ends of pipeline i; S_i is the impedance of pipeline I, including local resistance loss and on-way resistance loss; $H_{p,i}(n_i, Q_i)$ is the head of the water pump, which can be obtained from the head characteristic curve; $\phi_i(x_i)$ is the impedance of the valve, which is related to the valve opening x_i.

The vector form of Eq. (6.17) can be obtained to obtain a set of non-steady-state hydraulic equations for multiple pipelines:

$$K_f\dot{Q} = -(S + \phi)|\tilde{Q}|^d\tilde{Q} + H_p + \Delta\tilde{P}. \tag{6.19}$$

Combining graph theory and the transformation relationship between branches and leaves [8], the above equations can be combined to derive the non-steady-state hydraulic model of the entire pipeline network:

$$\begin{aligned}\dot{Q}_b = &-C_{r,b}^T\left(C_rK_fC_r^T\right)^{-1}C_r(S + \phi)\left|C_r^T\left(C_{r,b}^T\right)^{-1}\tilde{Q}_b\right|^d C_r^T\left(C_{r,b}^T\right)^{-1}\tilde{Q}_b\\ &+ C_{r,b}^T\left(C_rK_fC_r^T\right)^{-1}C_rH_p.\end{aligned} \tag{6.20}$$

When the initial flow rate of the fluid in the pipeline network is given, the dynamic distribution of the flow rate in the pipeline network after the adjustment of the water pump and water valve can be solved in conjunction with Eq. (6.20), and the time-delay characteristics of the water flow in the pipeline network can be described. For the nonlinear ordinary differential equation group, such as the above equation, the Euler method or the Runge–Kutta method can be used for solution. The Euler method has a relatively fast solution speed, but its solution accuracy is relatively low. The following is the solution process of the fourth-order Runge–Kutta method, defined as the iterative equation $f(\tilde{Q}_b)$:

$$f(\tilde{Q}_b) = -C_{r,b}^T\left(C_rK_fC_r^T\right)^{-1}C_r(S + \phi)\left|C_r^T\left(C_{r,b}^T\right)^{-1}\tilde{Q}_b\right|^d C_r^T\left(C_{r,b}^T\right)^{-1}\tilde{Q}_b$$

$$+ C_{r,b}^{T}\left(C_r K_f C_r^{T}\right)^{-1} C_r H_p. \tag{6.21}$$

Assuming that the branch initial water flow is $\tilde{Q}_b^0$, and the time interval is h, the flow rate of the branch in the $k + 1$ step can be expressed as:

$$\tilde{Q}_b^{k+1} = \tilde{Q}_b^k + \frac{h}{6}(k_1 + 2k_2 + 2k_3 + k_4). \tag{6.22}$$

The termination condition of the iteration can be set according to the precision requirement of the calculation result, where

$$k_1 = f\left(\tilde{Q}_b^k\right), \tag{6.23}$$

$$k_2 = f\left(\tilde{Q}_b^k + \frac{h}{2}k_1\right), \tag{6.24}$$

$$k_3 = f\left(\tilde{Q}_b^k + \frac{h}{2}k_2\right), \tag{6.25}$$

$$k_4 = f\left(\tilde{Q}_b^k + hk_3\right). \tag{6.26}$$

By solving Eq. (6.20), the hourly flow rate changes of the branches in the pipeline network can be obtained, and then combined with the transformation relationship between branches and leaves to obtain the hourly flow rate changes of the branches in the pipeline network.

6.1.3 Quantification of the Hysteresis Characteristics

In practical applications, the time required for the deviation between the output and the steady-state value to reach $\pm$ 5% or $\pm$ 2% is taken as the hydraulic response time. The hydraulic dynamic response time $t_{\mathrm{r,t}}$ of the water supply network can be defined as the time required for the temperature deviation between the supply water temperature at the cold source (or heat source) and the inlet temperature of all end points to reach within $\pm$ 2% after the start of a step change.

According to the pipeline thermal dynamic response model (2.2), the thermal dynamic response model of a certain pipe segment i in the network can be described as:

$$\tilde{T}_{t,i} + \frac{Q_i}{A_i}\tilde{T}_{x,i} = K_i(T_s - T_i). \tag{6.27}$$

where $K_i = \frac{1}{\rho c_p A_i R}$; Q_i is the fluid volume flow rate in pipe segment i; T_i is the temperature distribution in pipe segment i; $\tilde{T}_{x,i}$ is the partial derivative of the fluid temperature element in pipe segment i with respect to t; $\tilde{T}_{x,i}$ is the partial derivative of the fluid temperature element in pipe segment i with respect to x.

Combining the hydraulic model of the pipeline network, the change in flow rate Q_i in pipe i at time t can be obtained, and therefore, the thermal dynamic model of pipe i in the network can also be solved using the characteristic line method.

Using graph theory, the water supply pipeline network can be abstracted as a spanning tree, with the node at the cold source defined as the root node and the nodes at the end points defined as the leaf nodes. The temperature of the fluid when flowing into and out of each subnode can be considered as continuously changing. Therefore, the thermal characteristics of the water supply pipeline network can be obtained from the flow rates of each pipe segment, the initial temperature distribution, and the temperature changes at the inlet, thus obtaining the lag response time of the water supply pipeline network.

The thermal dynamic response time of the return water pipeline network can be defined as the time required for the deviation between the return water temperature at the least favorable end point and the new steady-state value at the cold source (or heat source) to reach $\pm$ 2% after the start of a step change. The thermal dynamic response time of the return water at a certain end point is the time required for the deviation between the return water temperature at that end point and the new steady-state value at the cold source (or heat source) to reach $\pm$ 2% after the start of a step change.

For the return water pipeline network, due to the presence of branch confluence, there is a mixing phenomenon when the fluid flows into and out of the nodes. As shown in Fig. 6.3, the temperature change at the nodes of the return water pipeline network can be established based on the law of energy conservation [9–11]:

$$\sum_i \left(Q_i T_{i,N}\right) = T_{o,0} \sum_i Q_i. \tag{6.28}$$

where the subscript i is the number of the incoming branch, o is the number of the main road after the branch confluence, N represents the control volume at the end of the pipe segment, and 0 represents the starting control volume of the pipe segment. This can be used to obtain the temperature change at the entrance of the main road after the convergence of the branch road. Combined with Eq. (6.28), the thermal dynamic response characteristics of the return water pipe network can be obtained. Then, combined with the thermal dynamic response model of pipeline i, the quality parameter lag time of the return water pipe network can be obtained.

For the hydraulic dynamic response time, the time required for the flow to reach 95% of the new steady-state value (Q_s) from the occurrence of a step change in the operating condition is defined as the hydraulic dynamic response time of the pipeline, denoted as $t_{\mathrm{r,h}}$. For a first-order ordinary differential equation as shown in Eq. (2.14), if:

$$\frac{2f}{\pi D^3} = \mathrm{a}, \quad \frac{\pi D^2}{4\rho L}\Delta p_1 = \mathrm{b}. \tag{6.29}$$

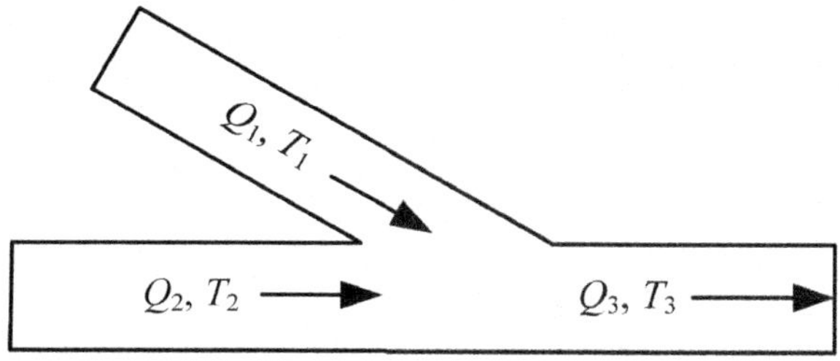

Fig. 6.3 Schematic diagram of pipe network branch confluence

The following can be obtained.

$$Q = c \cdot \tan h\left(\sqrt{ab} \cdot t + \theta\right). \tag{6.30}$$

The above equation is the non-steady-state hydraulic characteristic model of the pipeline, describing the change in flow rate over time when the pipeline fluid starts from the initial state (Q_0, Δp_0) and the pressure difference steps to Δp_1. Among them,

$$\theta = \frac{1}{2}\ln\left(\frac{c+Q_0}{c-Q_0}\right), \quad c = \sqrt{\frac{b}{a}} = \sqrt{\frac{\pi^2 D^5 \Delta p_1}{8f\rho L}}. \tag{6.31}$$

When $t \to \to +\infty$, the hyperbolic tangent function $\tanh t \to \to 1$, at which point

$$\lim_{t\to\infty} Q = c = \sqrt{\frac{\pi^2 D^5}{8f\rho L}\Delta p_1}. \tag{6.32}$$

According to the definition of hydraulic dynamic response time:

$$t_{\mathrm{r,h}} = \frac{1}{\sqrt{ab}}\left(\frac{1}{2}\ln\frac{1+0.95}{1-0.95} - \theta\right) = \sqrt{\frac{2\rho DL}{f\Delta p_1}}\left(\frac{1}{2}\ln 39 - \theta\right). \tag{6.33}$$

where f is the friction loss coefficient along the route, related to the Reynolds number, pipe wall roughness, and pipe diameter; the friction coefficient along the route in laminar flow (Re < 2320) can be approximately obtained by the following equation:

$$f = \frac{64}{\mathrm{Re}}. \tag{6.34}$$

The friction coefficient f in turbulent flow ($\mathrm{Re} \geq 2320$) can be calculated by the empirical formula of Haaland [12]:

$$\frac{1}{\sqrt{f}} = -1.8\lg\left[\frac{6.9}{\mathrm{Re}} + \left(\frac{e}{3.7D}\right)^{1.11}\right]. \tag{6.35}$$

where e is the absolute roughness of the pipe, related to the pipe material; Re is the Reynolds number of the fluid inside the pipe.

Equation (6.33) is the quantitative relationship of the lag time of the quantity parameter, and for different operating conditions and pipeline structures, the lag response time of the quantity parameter under the current operating conditions can be obtained.

6.2 Nonlinear Characteristics of the System Components

The previous section studied the model and simulation of the time-delay response characteristics of system parameters and explored the influence scale of the delay time of temperature and flow rate in the pipe network. This section will study the coupling relationship among different components. For the typical configurations of the HVAC water systems illustrated in Fig. 3.3, the optimization problem for the system in subplot (b) only needs to consider the load distribution optimization problem among the chillers. For the system configuration shown in subplot (a), the optimization problem is more complex, requiring consideration of the distribution optimization problem among parallel components and the diffusion optimization problem among different type components. Commonly, the optimization method for the system shown in subplot (a) can also be adopted to solve the optimization problem of the system shown in subplot (b). Therefore, this following research will mainly focus on the system configuration shown in subplot (a).

The energy efficiency of the HVAC water system is closely related to its own operating parameters, but due to the interactions among components, the operating parameters of the components present certain coupling characteristic. According to the component types, this coupling characteristic can be divided into the coupling relationship among different types of components and the coupling relationship among parallel components.

6.3 Coupling Relationship Among Different Type Components

From the spatial perspective, the coupling characteristics of the HVAC water system mainly exist among adjacent components, such as coupling characteristics between AHU and the chilled water pump, chilled water pump and chiller, chiller and cooling water pump, and cooling water pump and cooling tower. The following will formulate these coupling relationships.

(1) Coupling relationship between AHU and chilled water pump

There is a heat exchange process on the air side and water side of the AHU coil. When the mass-heat parameters on the air side change, it will affect the mass-heat parameters on the water side. If it is assumed that all the heat exchange of the AHU is used for cooling all the conditioned rooms, then its heat exchange process can be represented as Eq. (6.36).

$$\sum_{l=1}^{N_{\text{room}}} Q_{k,l} = \frac{e_{1,k} \cdot m_{\text{sa},k}^{e_{3,k}} \cdot \left(T_{\text{ma},k} - T_{\text{E,o}}\right)}{1 + e_{2,k} \cdot \left(m_{\text{sa},k}/m_{\text{chw},k}\right)^{e_{3,k}}}. \tag{6.36}$$

In the equation, e_1–e_3 are model fitting coefficients, which can be obtained through operation data identification; $Q_{k,l}$ is the cooling load of the lth room responsible by the kth AHU; T_{ma} is the mixed air temperature; $T_{\text{E,o}}$ is the coil inlet water temperature, scilicet the evaporator outlet temperature. In practical applications, the cooling load demand of different AHUs can be determined based on the actual demand of each conditioned room.

It can be seen that under the same operating conditions (Q, T_{ma}, and $T_{\text{E,o}}$ are the same), there is a coupling relationship between m_{sa} and m_{chw}. And these two parameters will affect the energy consumption of the AHU fan and the chilled water pump, respectively. For example, when the supply air volume of the AHU increases, its water demand will decrease, at this time the system will reduce the water flow by adjusting the AHU water valve to reduce the flow rate of the chilled water pump, thereby affecting the flow rate and head of the chilled water pump, and thus affecting the power of the chilled water pump. Therefore, under the same operating conditions, there is an optimal m_{sa} and m_{chw} that can optimize the energy consumption of the AHU fan and the chilled water pump.

(2) Coupling relationship between chilled water pump and chiller

The relationship between the chilled water pump and the chiller can be expressed by the law of energy conservation as shown in Eq. (6.37), namely the cooling capacity of the chiller is equal to the heat exchange capacity on the chilled water side.

$$\sum_{i=1}^{N_{\text{ch}}} Q_{\text{ch},i} = \left(\sum_{k=1}^{N_{\text{coil}}} m_{\text{chw},k}\right) c_{p,w}\left(T_{E,i} - T_{E,o}\right). \tag{6.37}$$

In the equation, Q_{ch} is the cooling capacity of the chiller unit, if it is assumed that all the cooling capacity of the chillers is used for cooling the conditioned rooms, it can be represented by Eq. (6.38); $c_{\text{p,w}}$ is the specific heat capacity of water at constant pressure; $T_{\text{E,i}}$ and $T_{\text{E,o}}$ are the inlet and outlet water temperatures of the evaporator, respectively.

$$\sum_{i=1}^{N_{\text{ch}}} Q_{\text{ch},i} = \sum_{k=1}^{N_{\text{coil}}} \sum_{l=1}^{N_{\text{room}}} Q_{k,l}. \tag{6.38}$$

It can be seen that under the same operating conditions (Q_{ch} and $T_{E,i}$ are the same), there is a coupling relationship between m_{chw} and $T_{E,o}$. For example, the smaller $T_{E,o}$, the smaller m_{chw}, and the changes in m_{chw} and $T_{E,o}$ will affect the energy consumption of the chilled water pump and the chiller, respectively. Therefore, under the same operating conditions, there is an optimal m_{chw} and $T_{E,o}$ that can optimize the energy consumption of the chilled water pump and the chiller.

(3) Coupling relationship between chiller and cooling water pump

For the relationship between chiller and cooling water pump, if it is assumed that all the heat converted from the compressor shaft power is taken away by the cooling water, the coupling relationship between the two can also be established based on the law of energy conservation, as shown in Eqs. (6.39) and (6.40).

$$P_{ch} + \sum_{i=1}^{N_{ch}} Q_{ch,i} = m_{cw} \cdot c_{p,w} \cdot \Delta T_C, \tag{6.39}$$

$$\Delta T_C = T_{C,o} - T_{C,i}. \tag{6.40}$$

In the equations, $T_{C,i}$ and $T_{C,o}$ are the inlet and outlet water temperatures of the condenser, and P_{ch} is the power of the chiller. It can be seen that under the same operating conditions, there is a coupling relationship between m_{cw} and $T_{C,o}$, and the changes in m_{cw} and $T_{C,o}$ will affect the energy consumption of the cooling water pump and the chiller, respectively. Therefore, under the same operating conditions, there is an optimal m_{cw} and $T_{C,o}$ that can optimize the energy consumption of the cooling water pump and the chiller.

(4) Coupling relationship between cooling water pumps and cooling towers

According to the law of energy conservation, the heat carried by the cooling water is equal to the heat dissipation of the cooling tower. Referring to the simplified heat dissipation model of the cooling tower, the coupling relationship between the cooling water pump and the cooling tower can be described as Eq. (6.41).

$$P_{ch} + \sum_{i=1}^{N_{ch}} Q_{ch,i} = \sum_{n=1}^{N_{ct}} \left(\frac{g_{1,n} \cdot m_{cw,n}^{g_{3,n}} \cdot \left(T_{C,i} - T_{wb,od}\right)}{1 + g_{2,n} \cdot \left(m_{cw,n}/m_{a,n}\right)^{g_{3,n}}} \right). \tag{6.41}$$

In this equation, g_1–g_3 are model fitting coefficients, which can be obtained through operation data identification; $T_{wb,od}$ is the outdoor wet bulb temperature, and $m_{a,n}$ and $m_{cw,n}$ are the air volume and water flow of the nth cooling tower, respectively.

It can be seen that under the same operating conditions, there is a coupling relationship between m_a and m_{cw}, and the changes in m_a and m_{cw} will, respectively, affect the energy consumption of the cooling tower fan and the cooling water pump. Therefore, under the same operating conditions, there is an optimal m_a and m_{cw} that can optimize the energy consumption of the cooling tower fan and the cooling water pump.

6.3.1 Coupling Relationship Among Parallel Components

The coupling relationship among parallel components mainly manifests as the problem of how to allocate the supply resources among parallel components with considering the energy efficiency trade-offs. Since components, such as chillers, water pumps, and cooling tower fans, all exhibit obvious nonlinear characteristics during operation, there exists an optimal energy efficiency point. That is, under the same operating conditions, there is an optimal supply resources that can make the total energy efficiency of the parallel components optimal. Therefore, it is necessary to balance the supply resources between parallel components and find the global optimal energy efficiency point based on the operating characteristics of each component.

For example, taking the water pump as an example, Fig. 6.4 shows a performance curve diagram of a variable frequency water pump. In which, the blue curve represents the head-flow characteristic curve, the black curve represents the efficiency-flow curve, the red curve represents the pipeline impedance characteristic curve, and the green curve represents the water pump operating curve based on the pressure difference ΔP control. It can be learnt from the figure that if the conventional control strategy is used to control the number of pumps, when the flow rate of the pump reaches the intersection of the operating curve and the characteristic curve (A_0 and B_0), an additional pump should be started to meet the control requirements, and the efficiency of the pump at this time is not optimal, and the pump has entered a low energy efficiency operating range. In comparison, if the switching point is selected as the operating point corresponding to the intersection of the efficiency curve (A^* and B^*), the number of pumps can be timely controlled according to the performance characteristics of each water pump, thereby ensuring the overall operating efficiency of the water pump is optimal.

Therefore, for multiple parallel water pumps, it is necessary to balance the flow rate between the water pumps, that is, how to reasonably allocate the water pump flow rate, so that the overall operating energy efficiency of the parallel water pumps reaches the optimal level. Herein, the water flow among parallel components will be allocated according to the proportion of the component rated values in the total sum of all operating component rated values, as shown in Eqs. (6.42) and (6.43), which represent the allocation of the chilled water flow (m_{chw}) among parallel chilled water pumps and the allocation of the cooling water flow (m_{cw}) among parallel cooling water pumps, respectively, where, N represents the component ON/OFF status, and the subscript "nom" represents the rated value of the component parameters.

$$m_{\text{chw},j} = \frac{m_{\text{chw,nom},j} \cdot N_{d,\text{chwp},j}}{\sum_{jj=1}^{N_{\text{chwp}}} \left(m_{\text{chw,nom},jj} \cdot N_{d,\text{chwp},jj}\right)} \cdot \left(\sum_{k=1}^{N_{\text{coil}}} m_{\text{chw},k}\right), \tag{6.42}$$

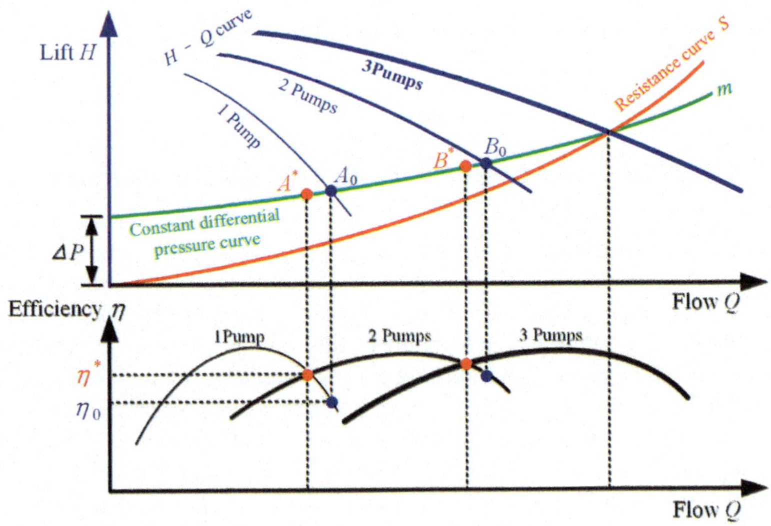

Fig. 6.4 Example of pump performance curve

$$m_{\text{cw},m} = \frac{m_{\text{cw,nom},m} \cdot N_{d,\text{cwp},m}}{\sum_{mm=1}^{N_{\text{cwp}}} \left(m_{\text{cw,nom},mm} \cdot N_{d,\text{cwp},mm}\right)} \cdot m_{\text{cw}}. \tag{6.43}$$

For the chillers, the operating characteristics are closely related to the system cooling load, as shown in Fig. 6.5. Therefore, for a given cooling load demand, it is necessary to allocate the cooling load among chillers based on the operating characteristics of each chiller unit, so that the total energy efficiency of the chillers is optimal while meeting the cooling load demand, as shown in Eq. (6.44), which represents the allocation of the cooling load (Q_{ch}) among parallel chillers.

$$Q_{\text{ch},i} = \frac{Q_{\text{ch,nom},i} \cdot N_{d,\text{ch},i}}{\sum_{ii=1}^{N_{\text{ch}}} \left(Q_{\text{ch,nom},ii} \cdot N_{d,\text{ch},ii}\right)} \cdot \left(\sum_{k=1}^{N_{\text{coil}}} \sum_{l=1}^{N_{\text{room}}} Q_{k,l}\right). \tag{6.44}$$

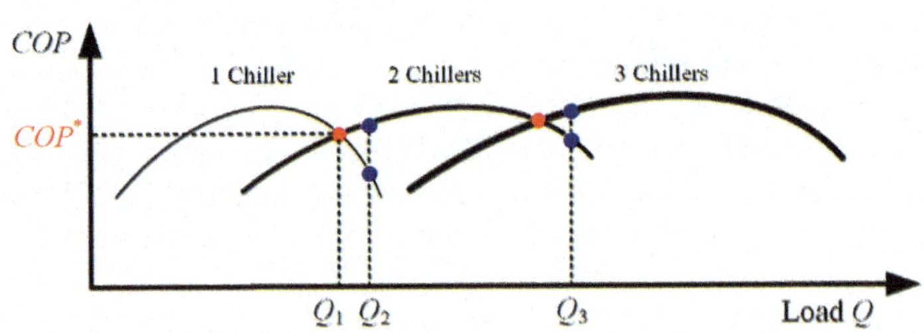

Fig. 6.5 Example of chiller performance curve

Since the operating characteristics of the fans are similar to those of the pumps, there is also the aforementioned energy efficiency trade-off problem among the cooling tower fans. Therefore, parallel cooling towers need to allocate the air volume between the cooling tower fans, and similarly, the cooling water flow also needs to be allocated among cooling towers. As shown in Eqs. (6.45) and (6.46), which represent the allocation of the air volume (m_{a}) and the cooling water flow (m_{cw}) among parallel cooling towers, respectively.

$$m_{a,n} = \frac{m_{a,\mathrm{nom},n} \cdot N_{d,\mathrm{ct},n}}{\sum_{nn=1}^{N_{\mathrm{ct}}} \left(m_{a,\mathrm{nom},nn} \cdot N_{d,\mathrm{ct},nn} \right)} \cdot m_{a}, \tag{6.45}$$

$$m_{\mathrm{cw},n} = \frac{m_{\mathrm{cw},\mathrm{nom},n} \cdot N_{d,\mathrm{ct},n}}{\sum_{nn=1}^{N_{\mathrm{ct}}} \left(m_{\mathrm{cw},\mathrm{nom},nn} \cdot N_{d,\mathrm{ct},nn} \right)} \cdot m_{\mathrm{cw}}. \tag{6.46}$$

Based on the above description, the energy efficiency trade-off issue between parallel components can be understood as the following two cases:

(a) If the performance of each component in the parallel components is identical, then the energy efficiency trade-off among components is a problem of optimizing the number of operating units, and it is only necessary to determine the number of components to be turned on, and the supply resources will be evenly distributed among the operating components;
(b) If the parallel components have different sizes or performance, then this problem is an optimization problem of ON/OFF sequence, and it is necessary to determine the optimal combination of component ON/OFF states, and the supply resources for each component will be allocated according to its rated value as a proportion of the total rated value of all operating components.

6.4 Simulation Research of the Hysteresis Characteristics

6.4.1 Simulation Methods

In order to clearly and intuitively reflect the impact of the coupling relationship between various factors on the time lag of parameters, simulation studies will be carried out on the effects of various factors using the full orthogonal and pairwise orthogonal experimental methods.

The full orthogonal experimental method tests every combination of all levels of different factors selected, in order to obtain comprehensive experimental results. Its advantage is that it can accurately reflect the influence of each factor on the research target, while maximizing the scale of the interaction between the factors. Its disadvantage is that many experiments need to be conducted, and when multiple factors and levels are involved in the study, the number of experiments often increases geometrically.

Compared with the full orthogonal experimental method, the pairwise orthogonal experimental method is a more efficient experimental design method that seeks the optimal level combination. Its characteristic is that the overall experimental situation can be obtained by analyzing only part of the experimental results. The experimental combination is based on the principle of orthogonality and the principle of balanced distribution of experimental points, and some representative experimental points are selected from the full orthogonal combination. The orthogonal table required for pairwise orthogonal experiments is calculated using the pairwise algorithm proposed by American psychologist L. L. Thurstone in 1927, which is a scientific and efficient experimental method widely used in multi-factor experimental research. If the research object is related to n factors and each factor has m levels, a total of mn experiments are required to complete the full orthogonal experiment; only $m \times n$ tests are required for pairwise orthogonal experiments.

Based on the characteristics of the above two experimental methods, the full orthogonal experimental method can be used to obtain a comprehensive and universal research result for the time-delay response characteristics of pipelines under different volumes and operating conditions. For more specific research on time-delay response characteristics, the pairwise orthogonal experimental method can be used for case analysis of the time-delay response characteristics of pipelines under different volumes and operating conditions, and to further study the scale of the influence of various factors on time-delay characteristics. The parameter settings of the simulation research are listed in Table 6.1.

Table 6.1 Parameter setting for thermal dynamic response simulation

Name	Notation	Value
Specific heat	c_p	4200 J/(kg K)
Density	ρ	960 kg/m^3
Fluid heat transfer coefficient	k_w	0.599 W/(m K)
Dynamic viscosity	μ_w	1.11×10^{-3} Pa s
Wall thermal conductivity	k_{12}	30.0 W/(m K)
Thermal conductivity of thermal insulation layer	k_{23}	0.03 W/(m K)
Insulation thickness	δ	Refer to GB50736-2012
Natural convection heat transfer coefficient of air	h_a	5.0 W/(m^2 K)
Environment temperature	T_s	26 °C
Initial fluid temperature	T_0	26 °C, uniform distribution
Supply water temperature	$T_0{}^n$	7 °C, constant

6.4.2 *Impacts of Pipe Parameters on Temperature Hysteresis*

According to the above definition and operating conditions, Fig. 6.6 shows the orthogonal simulation results of the thermal dynamic response characteristics in the water pipe. The simulation involves factors including pipe diameter and length, and the flow rate inside the pipe is 1.2 m/s.

The curved surface in the simulation results reflects the differences in the response time of the pipeline under the same operating conditions and different combinations of pipe diameter and length. The projection of the surface on the *L*–*T* plane and the *D*–*T* plane represents the thermal response time curves under the same pipe diameter and the response time curves under the same pipe length, respectively. It can be seen from the simulation results that the orthogonal simulation results are approximately a plane, indicating that the influence of the pipe diameter on the thermal response time is very slight, while the pipe length has a significant impact on the dynamic thermal response time of the pipeline, and the longer the pipeline length, the longer the thermal response time of the pipeline. Therefore, under the same operating conditions, the thermal dynamic response time of the pipeline is positively correlated with the pipe length and can be considered unrelated to the pipe diameter.

Figure 6.7 compares the orthogonal simulation results of the thermal dynamic response characteristics of the water pipe at different flow rates. It can be seen from subplot (a) that the higher the flow rate inside the pipe, the shorter the thermal response time; and as the flow rate increases, the influence on the thermal response time gradually decreases. The curved surface in subplot (b) reflects the differences in the thermal response time of the pipeline under the same operating conditions and different combinations of flow rate and pipe length; the projection of the surface on the *L*–*T* plane and the *v*–*T* plane represents the curve clusters of the thermal response time with the pipe length under the same flow rate and the curve clusters of the response time with the flow rate under the same pipe length, respectively. The simulation results show that both the pipe length and the flow rate have a significant impact on the dynamic thermal response time of the pipeline, and the longer the

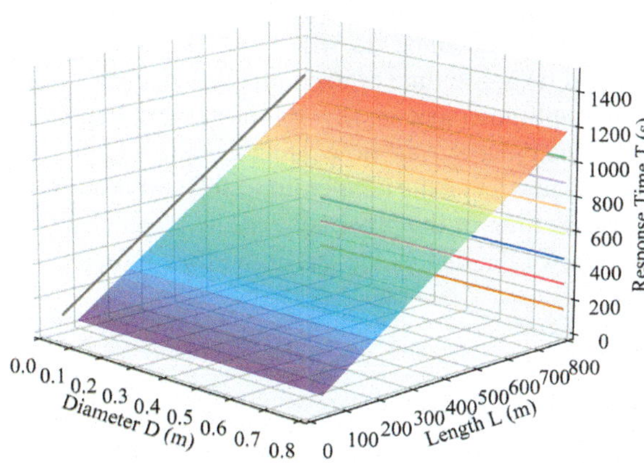

Fig. 6.6 Orthogonal simulation results of thermal response characteristics in different pipes (redrawn from Ref. [4], copyright 2022, with permission from ELSEVIER)

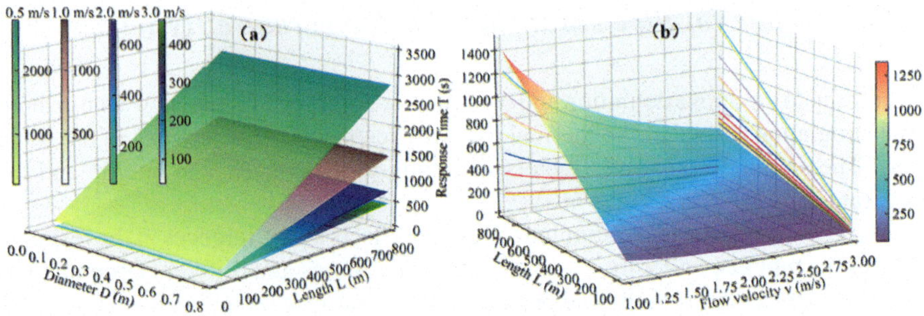

Fig. 6.7 Simulation results of thermal response characteristics with different flowrate (redrawn from Ref. [4], copyright 2022, with permission from ELSEVIER)

pipeline length and the smaller the flow rate, the longer the dynamic thermal response time of the pipeline; from the projection of the surface on the two planes, it can be seen that the dynamic thermal response time of the pipeline is linearly proportional to the pipe length and inversely proportional to the flow rate.

6.4.3 Impacts of Temperature Step on Temperature Hysteresis

The amplitude of temperature step is defined as the absolute value of the difference between the initial temperature of the pipeline (uniform distribution) and the inlet supply temperature. It can be divided into two situations. The first situation is that the supply temperature is constant, but the initial temperature of the fluid inside the pipeline is different. The second situation is that the initial temperature of the fluid inside the pipeline is the constant, but the supply temperature is different. The pair-wise orthogonal simulation results of the thermal dynamic response characteristics in the two situations are shown in Figs. 6.8 and 6.9, respectively, with a uniform pipe length of 200 m.

From the simulation results, it can be seen that the thermal response time in both situations can be classified into three categories according to the flow rate. The thermal response time of each pipeline in each category varies with the temperature step amplitude, specifically, the larger the step amplitude, the longer the response time, and the lower the flow rate, the more significant the difference. In conclusion, it can be known that the temperature step amplitude has a certain impact on the thermal response time of the pipeline, and the impact effect is inversely proportional to the fluid flow rate, that is, the larger the flow rate, the smaller the impact scale.

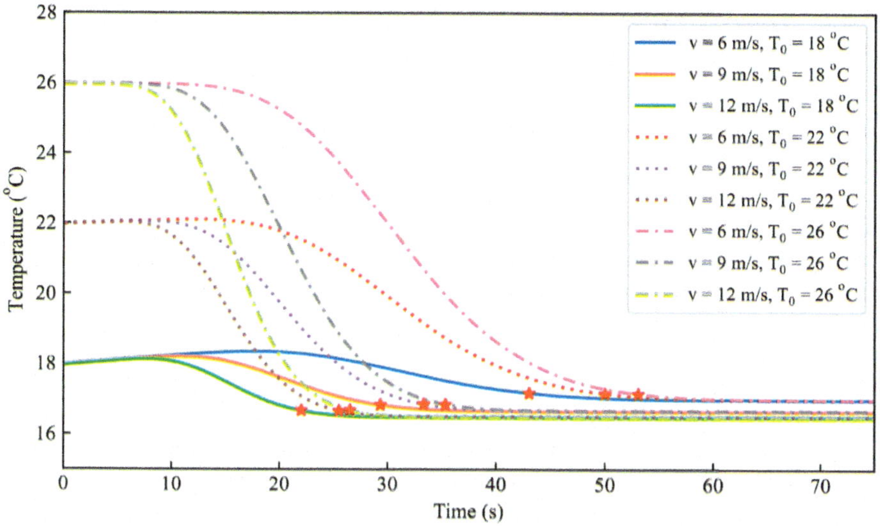

Fig. 6.8 Pairwise orthogonal simulation of thermal response characteristics in water pipes with different initial temperatures (redrawn from Ref. [4], copyright 2022, with permission from ELSEVIER)

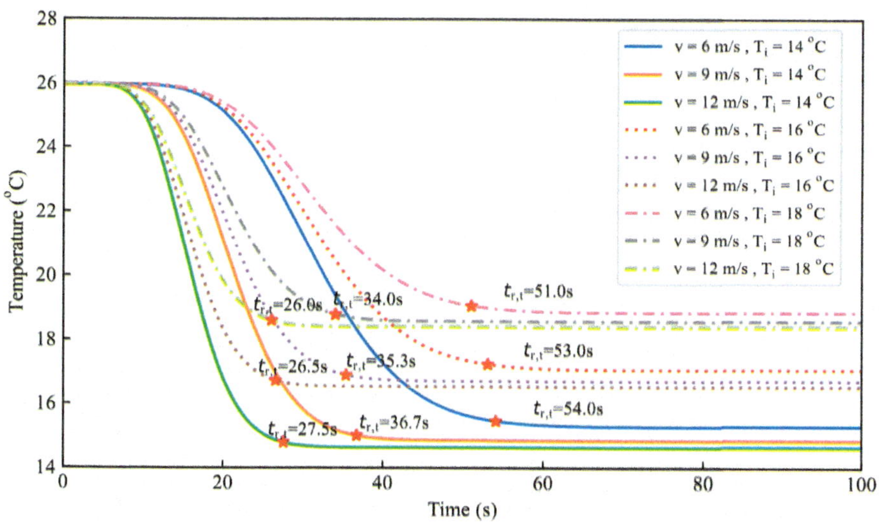

Fig. 6.9 Pairwise orthogonal simulation of thermal response characteristics in water pipes with different inlet temperatures (redrawn from Ref. [4], copyright 2022, with permission from ELSEVIER)

6.4.4 Impacts of Pipe Parameters on Water Flow Hysteresis

During the simulation, the default pipe material is seamless steel pipe, and the absolute roughness is taken as 0.03 mm. Figure 6.10 shows the orthogonal simulation results of the hydraulic response time of the water pipe. Among them, subplot (a) shows the full orthogonal simulation results when the pressure difference at both ends of the pipeline changes from 0 to 20 kPa, and the factors involved in the simulation include pipe diameter and pipe length. The surface in the simulation results reflects the differences in the hydraulic response time of the pipeline under different combinations of pipe diameters and lengths under the same operating conditions. The projection of the surface on the *L–T* plane and the *D–T* plane shows the curves of the hydraulic response time under the same pipe diameter and the curves of the response time under the same pipe length, respectively. It can be seen from the simulation results that both the pipe length and diameter have a significant impact on the hydraulic dynamic response time, and the longer the pipe length and the larger the pipe diameter, the longer the hydraulic dynamic response time of the pipeline; and the projection of the surface on the two planes can reflect the differences in the hydraulic dynamic response time of different pipe sizes, and the large size leads to the significant difference.

Subplot (b) shows the orthogonal simulation results of the pressure difference of the pipeline when the pressure difference ranges from 0 kPa to 10, 20, 30, and 40 kPa, respectively. It can be seen that when the pressure difference is 10 kPa, the response time of the pipeline is the longest, and when the pressure difference is 40 kPa, the response time of the pipeline is the shortest, about half of that at 10 kPa. Therefore, under the same pipe diameter and length, the larger the pressure difference at both ends of the pipeline, the shorter the hydraulic dynamic response time; and with the increase of the pressure difference, the gap between adjacent surfaces gradually

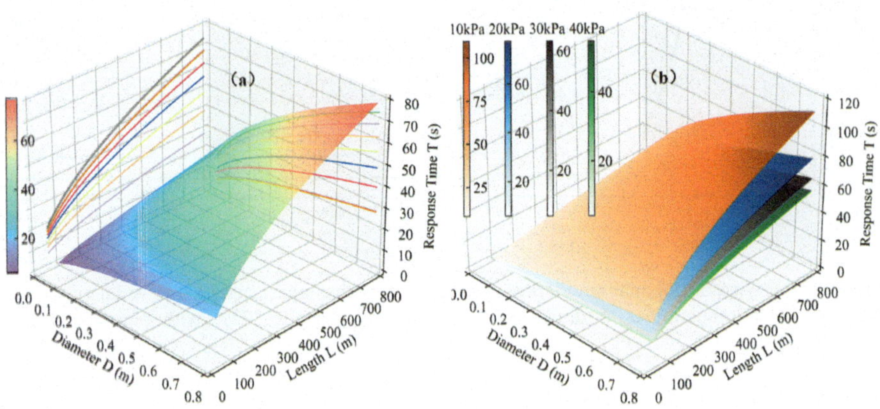

Fig. 6.10 Orthogonal simulation results of hydraulic characteristics in water pipe (redrawn from Ref. [4], copyright 2022, with permission from ELSEVIER)

decreases, indicating that the effectiveness of the pressure difference on the hydraulic response characteristics of the pipeline gradually decreases with the increase of the pressure difference.

Based on the above full orthogonal simulation experiments, it can be concluded that the pipe diameter, pipe length, and pressure difference at both ends are the main factors affecting the time lag of water flow. The following will focus on these three factors, select three levels for each factor, and use the pairwise orthogonal simulation method for a more in-depth case comparison analysis. The simulation involves three factors, each factor takes three levels, and there are a total of 9 cases according to the pairwise algorithm.

Figure 6.11 shows the comparison results of the time lag of flow and flow rate for the 9 cases, and the red asterisks on the curves represent the points where the fluid reaches a steady state. From the comparison curve of flow, it can be seen that the curve with the smallest steady flow corresponds to the smallest pipe diameter of DN100, and the scatter distribution of response time can be classified into three categories according to the pipe diameter, among which the DN300 pipe corresponds to the largest steady flow, indicating that the pipe diameter plays a dominant role in the size of the steady flow. It can be known that the larger the pipe diameter, the shorter the pipe length, and the larger the pressure difference at both ends, the higher the steady flow that the fluid in the pipeline can reach. From the flow rate curve, it can be seen that the time scatter of steady flow rate can also be classified into three categories according to the pipe diameter, and the comparison between the categories shows that the larger the pipe diameter, the longer the time lag of water flow. The comparison between points within each category shows that under the same pipe diameter, the shorter the pipe length and the larger the pressure difference, the greater the flow rate after the fluid reaches a steady state.

6.5 Summary

This chapter investigates the application methods of spatiotemporal feature models in practical systems from the perspective of practical application. It proposes an online application method based on a dual-loop control architecture. Through experimental studies, the reliability of the spatiotemporal feature model and the online application method has been verified, effectively enhancing the feasibility of the model and method in practical applications. This chapter provides theoretical guidance and technical support for the online application of decoupling optimization control methods for HVAC water systems.

Based on the research content of this chapter, the following conclusions are drawn:

(1) The online application method of the spatiotemporal feature model of HVAC water systems based on the dual-loop control architecture can obtain real-time system status through the TCP/IP protocol. It dynamically predicts the state distribution of quality-mass parameters in the pipeline network based on

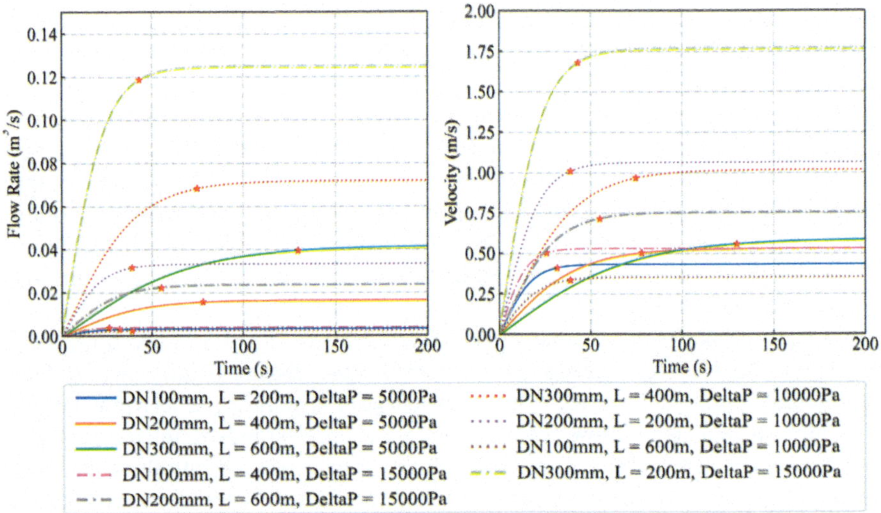

Fig. 6.11 Pairwise orthogonal simulation results of hydraulic characteristics in water pipe (redrawn from Ref. [4], copyright 2022, with permission from ELSEVIER)

the current system state and control actions. The average durations required for hydraulic and thermal calculations per iteration are 1.18 and 0.19 ms, respectively, meeting the computational requirements of online applications.

(2) The quality-mass parameter time-delay response characteristic model of the HVAC water system has high prediction accuracy. Experimental results show that the average R2 values of the quality parameter and quantity parameter time-delay response characteristic models are 0.98 and 0.96, respectively, meeting the control requirements for subsequent decoupling optimization of HVAC water systems.

(3) From the experimental results, it can be seen that there are differences in the time lag scale and steady-state value of the response to changes in supply water temperature to different end branches, which are more pronounced in HVAC water systems with higher spatial complexity. Therefore, in the optimization and control process of practical systems, research on decoupling optimization control methods based on spatiotemporal features needs to be conducted for different branches or branch control loops.

(4) Through the spatiotemporal feature analysis of the HVAC water system, the coupling and causality relationships between system quality-mass parameters, equipment, and control loops have been clarified. The importance of system spatiotemporal features on operational efficiency and optimization control effectiveness has been determined, providing a clear research direction for the implementation of decoupling optimization methods for practical HVAC water systems.

References

1. Duquette J, Rowe A, Wild P (2016) Thermal performance of a steady state physical pipe model for simulating district heating grids with variable flow. Appl Energy 178:383–393. https://doi.org/10.1016/j.apenergy.2016.06.092
2. Steer KCB, Wirth A, Halgamuge SK (2011) Control period selection for improved operating performance in district heating networks. Energy Build 43:605–613. https://doi.org/10.1016/j.enbuild.2010.10.029
3. Gabrielaitiene I, Bøhm B, Sundén B (2011) Dynamic temperature simulation in district heating systems in Denmark regarding pronounced transient behaviour. J Civ Eng Manag 17:79–87. https://doi.org/10.3846/13923730.2011.553936
4. Wang J, Zhao T (2022) Medium dynamic diffusion characteristics in the pipeline network of variable flow HVAC system towards online decoupling control oriented. Energy Build 258:111827. https://doi.org/10.1016/j.enbuild.2021.111827
5. Lax PD, Wendroff B (1962) On the stability of difference schemes. Commun Pure Appl Math 15:363–371. https://doi.org/10.1002/cpa.3160150401
6. Ravnik J, Škerget L, Hriberšek M (2004) The wavelet transform for BEM computational fluid dynamics. Eng Anal Bound Elem 28:1303–1314. https://doi.org/10.1016/j.enganabound.2004.05.002
7. Wylie EB, Streeter VL (1978) Fluid transients. McGraw-Hill International Book Co., New York
8. Fu XZ, Xiao YM (2018) Fluid network for transportation and distribution, 4th edn. China Architecture & Building Press, Beijing
9. Wang D, Zhi Y, Jia H, Hou K, Zhang S, Du W, Wang X, Fan M (2019) Optimal scheduling strategy of district integrated heat and power system with wind power and multiple energy stations considering thermal inertia of buildings under different heating regulation modes. Appl Energy 240:341–358. https://doi.org/10.1016/j.apenergy.2019.01.199
10. Wang H, Meng H (2018) Improved thermal transient modeling with new 3-order numerical solution for a district heating network with consideration of the pipe wall's thermal inertia. Energy 160:171–183. https://doi.org/10.1016/j.energy.2018.06.214
11. Wang Y, Shi K, Zheng X, You S, Zhang H, Zhu C, Li L, Wei S, Ding C, Wang N (2020) Thermo-hydraulic coupled analysis of meshed district heating networks based on improved breadth first search method. Energy 205:117950. https://doi.org/10.1016/j.energy.2020.117950
12. Çengel YA, Cimbala JM (2006) Fluid mechanics: fundamentals and applications. McGraw-Hill Higher Education, Boston

Chapter 7
Global Optimization Problem and Methods

The previous chapter conducted an in-depth study on the hysteresis characteristics of the medium in the pipeline network and the system nonlinear characteristics. The results show that both the temperature and flow rate of the medium exhibit hysteresis characteristics during the response process, which dynamically change with the system configuration and control actions, affecting the overall performance and control effectiveness. With the increasing demand for intelligent HVAC control systems, the periodic optimization is no longer sufficient to adapt to complex and variable operating conditions, often resulting in untimely or unnecessary optimization actions that impact system performance. Therefore, determining the optimal timing for system optimization and control during the process is crucial to enhance the adaptability of the variable flowrate HVAC system to operating conditions. This chapter will present a nonlinear optimization method by combining with medium hysteresis characteristics and system nonlinear characteristics, to enhance the overall stability of the HVAC water system and its adaptability to random operating conditions. Additionally, experiments will be conducted to compare the performance of the proposed method with the traditional method in actual systems, comprehensively evaluating the two methods in terms of decoupling effect, control effectiveness, and energy consumption.

T. Zhao et al., *Thermal Distribution Principle and Nonlinear Optimal Control for HVAC Water System*, Indoor Environment and Sustainable Building,
https://doi.org/10.1007/978-981-96-0377-0_7

7.1 Optimization Timing and Event-Driven Optimization Method

7.1.1 Optimization Timing Problem

For conventional optimization control, the activation mode of the optimization process typically employs a time-driven approach. In this method, the control loop setpoint of the system is periodically triggered for optimization based on a predetermined time interval. However, this triggering mechanism is somewhat mechanical and exhibits two notable drawbacks:

1. In the event of changes in operating conditions, such as fluctuations in outdoor temperature or significant personnel flow, the control system fails to adjust promptly, leading to a missed opportunity for optimal adjustment.
2. In scenarios where operating conditions remain relatively stable, the system still undergoes "unnecessary" optimization adjustments according to the preset cycle. This can adversely affect the stability of the system.

In the time-driven optimization (TDO) strategy, optimization is triggered based on "time", while in the event-driven optimization (EDO) strategy, the optimization trigger depends on "events" and is closely related to the system's operating state. Figure 7.1 shows the flowcharts of the two strategies. It can be seen that in the TDO strategy, the system is driven to optimize the setpoint periodically. In the EDO strategy, optimization only occurs when an "event" is triggered. If no "event" is triggered, the system will maintain the current state without any optimization. In comparison, the trigger mode of the EDO strategy is more flexible, allowing it to promptly capture feedback on the system's operational process and make necessary optimization adjustments. This can effectively mitigate the drawbacks of the TDO strategy, enhancing the overall operational efficiency and optimization effects of the system. Additionally, since the EDO strategy is only related to the system's state, it can better avoid the uncertainty of parameter delays.

7.1.2 Event-Driven Optimization Method

During the implementation phase of EDO strategy, a fixed time interval (e.g., 5 min or 10 min) is typically specified as the event recognition period to mitigate frequent event triggering. However, practical applications often face challenges due to the absence of standardized rules for defining event triggering intervals. The delay response characteristics of the working substance in various system volumes can result in issues, such as "untimely triggering" or "repeated triggering". The latter occurs when the current optimization process has not yet responded but has reached the triggering period, entering the next optimization cycle. These challenges frequently undermine the advantages of the EDO strategy. To overcome the limitations of both

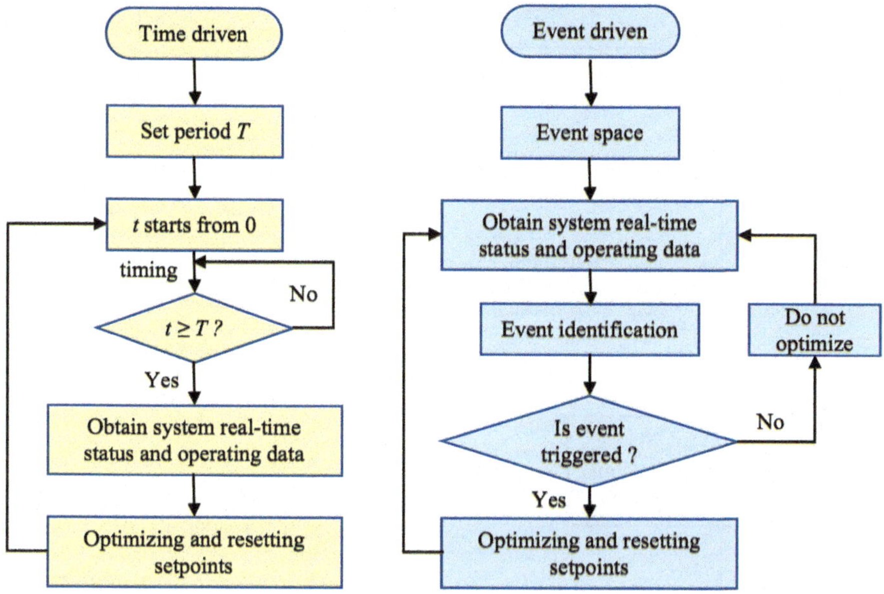

Fig. 7.1 Flowchart of two optimization strategies

TDO and EDO strategies and achieve a more intelligent and efficient optimization control method for the HVAC systems, this section presents a novel optimization strategy—the sequential event-driven (SEDO) optimization strategy.

The triggering process of the SEDO strategy is illustrated in Fig. 7.2, with events determined by monitoring real-time operating data and system conditions. The "preparation state" serves as a flag indicating whether event identification can proceed, effectively organizing the optimization process to ensure orderly progress. This arrangement prevents multiple optimization processes from coupling and interfering with each other. If the preparation state is "*OFF*", it indicates an ongoing optimization process, temporarily preventing the next optimization control. Upon completion of the optimization process, the preparation state transitions from "*OFF*" to "*ON*", signaling that the preparation state is active, and the next optimization process can be executed at any time. Moreover, in conjunction with previous research on dynamic propagation characteristics, the preparation state serves as the endpoint for the system's time-delay response. Entering the preparation state ("*ON*") indicates that the preceding dynamic adjustment has fully responded to its conclusion.

Figure 7.3 presents a comparison of the schematic diagrams of the sequential event-driven optimization (SEDO) and time-driven optimization (TDO) strategies. While both strategies share the same sampling period, the key distinction lies in the optimization period for the setpoint. In the TDO strategy, each period comprises three stages: trigger period, optimization period, and idle period. The operational sequence for the entire period is as follows: The optimization process is triggered periodically at fixed time intervals (T); subsequently, the system executes the optimization algorithm

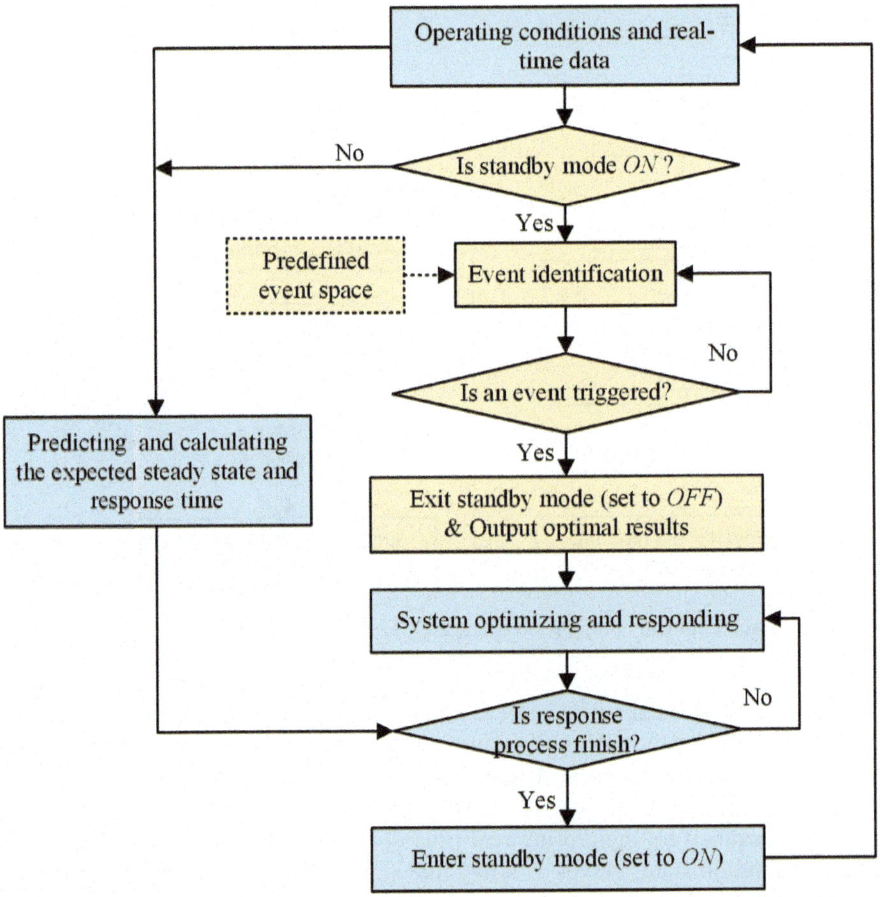

Fig. 7.2 Flowchart of SEDO strategy

based on the current state; finally, after providing the optimization results, it enters the idle period until the next trigger in the subsequent period.

In contrast to the TDO strategy's periodic trigger mechanism, the optimization period for the SEDO strategy varies based on demand response, enabling a more precise grasp of the optimization timing. The complete cycle of the SEDO strategy encompasses four stages: trigger period, optimization period, response period, and standby period. During the trigger period, the system identifies events based on the preset event space. Once it detects events meeting the conditions, it triggers optimization, changing the standby status from "*ON*" to "*OFF*". In the optimization period, the optimization algorithm calculates the optimal setpoint, which is then transmitted to the corresponding control loop. In the response period, each control loop tracks the new setpoint using the control algorithm. This process continues until the system achieves the control objective or the operating state stabilizes, ensuring

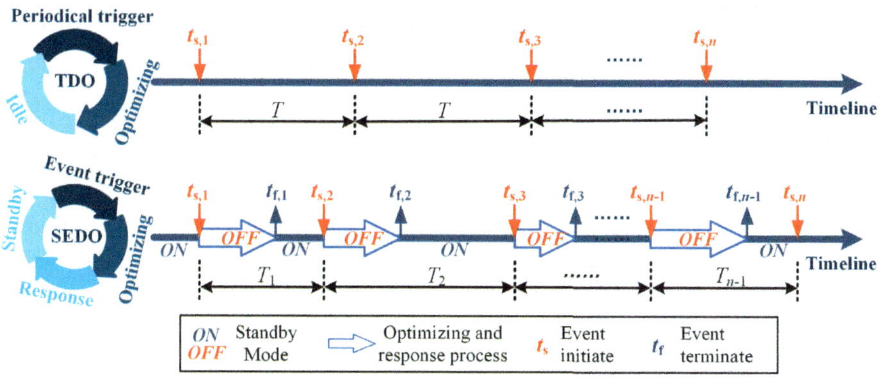

Fig. 7.3 Schematic contrast between TDO and SEDO strategies (redrawn from Ref. [1], copyright 2022, with permission from ELSEVIER)

that the quality parameter fluctuations caused by the control action complete the response from the "source" to the "end". Following the response, the standby status shifts from "*OFF*" to "*ON*", signifying that the next optimization can be performed at any time. The time lag scale is associated with the pipeline network structure, the current operating state of the system, and the control action. It can be utilized in conjunction with the response characteristics of the parameter time lag from previous research for prediction.

7.1.3 Event Space and Optimal Timing Determination

In the HVAC water system, the operational state of the system can be characterized through "events". Different "events" are defined to portray the reasonable fluctuation range of various state parameters (such as Part Load Ratio (PLR) and wet bulb temperature). Following the definition of "events" in the event-driven optimization (EDO) strategy, an effective event primarily encompasses three attributes: timestamp, state parameters, and judgment criteria [2]. These three attributes collectively establish the criteria for determining whether an event is triggered, with the event being considered triggered only when the state parameters align with the judgment criteria. The event space (E) is defined as a set composed of several events (e_i), and this can be expressed mathematically as Eq. (7.1) [3]:

$$E = \{e_i\}, \quad i \in R. \tag{7.1}$$

If "significant changes in parameters" are established as the criteria for triggering events, then the events corresponding to continuous variables and discrete variables can be defined according to Eqs. (7.2) and (7.3), respectively [2].

$$e_i = \{\langle M_s, Y_i^{\text{prev}}, Y_i^{\text{curr}}\rangle | M_s = ON \;\&|Y_i^{\text{prev}} - Y_i^{\text{curr}}| \geq \zeta_i\}, \tag{7.2}$$

$$e_i = \{\langle M_s, Y_i^{\text{prev}}, Y_i^{\text{curr}}\rangle | M_s = ON \;\& Y_i^{\text{prev}} \neq Y_i^{\text{curr}}\}. \tag{7.3}$$

In the equations, Y represents the vector of state parameters, ζ is the threshold corresponding to the state parameters, the subscript i denotes the event number, and the superscripts prev and curr indicate the state values at the previous moment and the current moment. M_s signifies the standby state, indicating that event determination can only occur when the standby state is "*ON*". Currently, the methods for selecting the event space and judgment criteria for the HVAC water system mainly include empirical methods and theoretical methods. Empirical methods are typically based on current standards or operational experience, such as those grounded in expert knowledge and those relying on prior knowledge. These methods can be employed to determine the event space and judgment criteria. However, this approach is limited to describing a specific range of events, making it challenging to incorporate the characteristics of the system itself to determine the event space. It also exhibits poor applicability to specific systems or non-standard operating conditions [2].

In contrast, theoretical methods, such as data mining technology [4], are more flexible, are specific, and have broader applicability. They possess a stronger ability to describe the event space.

7.1.4 Application of Hysteresis Characteristics

When it comes to optimizing the decoupling of complex systems, three main aspects are typically considered. Firstly, prioritize the main controlled parameters based on regulatory requirements, neglecting less influential parameters to simplify the control process. Secondly, for "single-to-multi" or "multi-to-single" input–output patterns, select the most correlated input–output parameters to establish multiple "one-to-one" loops, thereby reducing coupling between loops. Thirdly, a commonly used decoupling method involves designing a decoupling compensator $G_p(s)$ based on the control system characteristics. However, this method demands high accuracy in the controlled object model, making it challenging to apply in practical systems.

The SEDO strategy initiates the optimization process by addressing system decoupling from the first aspect, ensuring that the response of dominant parameters is prioritized. In practical applications, lag time serves as the basis for determining the dominant position. When both qualitative and quantitative parameter adjustments occur, the parameter with the longer lag time takes precedence in meeting its control requirements.

As depicted in Fig. 7.2, the sequential event-driven optimization (SEDO) strategy involves determining whether the current optimization process is fully responsive, based on the previously studied parameter lag response characteristics. The application process of the lag response characteristics in the SEDO strategy is illustrated in

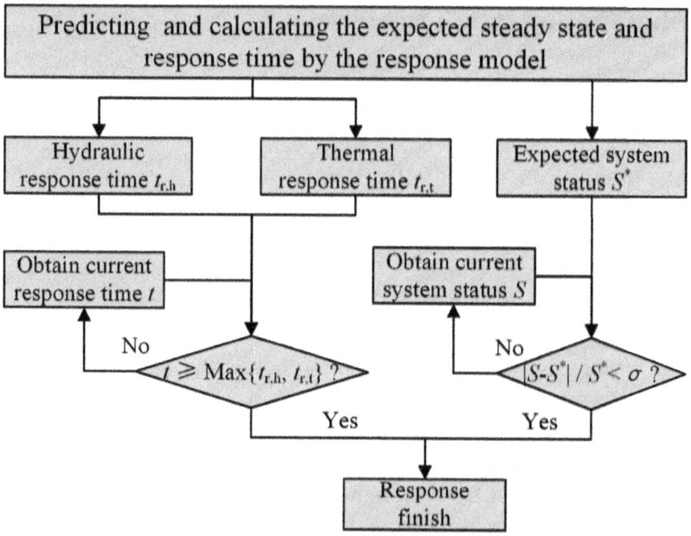

Fig. 7.4 Application of response characteristics in SEDO strategy

Fig. 7.4. Firstly, utilizing the lag response characteristic model of the system's qualitative and quantitative parameters, the required lag times ($t_{r,h}$ and $t_{r,t}$) and the expected parameter distribution state S^* after the control action can be calculated. Secondly, the duration t of the control action and the current system state S are assessed. If both the qualitative and quantitative parameters have responded within the duration t, or if the system state S is close to the expected state S^*, then it is considered that the qualitative and quantitative parameters have fully responded from the "source" to the "end" during this adjustment.

In conclusion, the optimization cycle of the SEDO strategy varies with demand response and is related to parameter lag time and the duration of the preparatory state. Simultaneously, the strategy fully considers the lag response characteristics of the qualitative and quantitative parameters within the pipeline network, which is highly significant for improving the coupling relationship. From the perspective of achieving decoupling of qualitative and quantitative parameters, the SEDO strategy can prioritize the control requirements of the more influential controlled parameters, ignoring the control requirements of the less influential ones, thereby simplifying the decoupling process. It is evident that the SEDO strategy can consider the coupling mechanism of the qualitative and quantitative parameters in the pipeline network, optimize and control the system at the appropriate time, and has the ability to achieve high-quality decoupling optimization control.

7.2 Global Optimization Problem

7.2.1 Mathematical Description

The global optimization goal of the HVAC water system is to minimize total power consumption while meeting cooling demand. Thus, the objective function is represented by Eq. (7.4).

$$f(X) = \min P_{\text{sys}} = \min\left(P_{\text{ch}} + P_{\text{chwp}} + P_{\text{cwp}} + P_{\text{ct}} + P_{\text{ahu}}\right). \tag{7.4}$$

Based on power consumption models and the energy efficiency trade-off mechanism, there are optimal setpoints for control loops and the ideal *ON/OFF* state for components to minimize the system's overall power consumption. The optimization process is expressed in Eq. (7.5). In this equation, $\{N_{\text{d, comp}}\}$ represents the collection of numbers for different operating components, such as chillers, chilled water pumps, cooling water pumps, and cooling towers. The optimization parameters $T_{\text{E,o}}$, H_{chwp}, ΔT_{C}, $T_{\text{C},i}$, T_{sa}, and H_{sa} are control loop setpoints. These can be managed and tracked by adjusting the chiller operating parameters, chilled water pump speed, cooling water pump speed, cooling tower fan speed, cooling coil valve opening, and AHU supply fan speed, respectively.

$$\begin{aligned}&\left(\left\{N^*_{\text{d,comp}}\right\}, T^*_{\text{E,o}}, H^*_{\text{chwp}}, \Delta T_{\text{C}^*}, T^*_{\text{C,i}}, T^*_{\text{sa}}, H^*_{\text{sa}}\right)\\ &\quad = \arg\min f\left(\left\{N_{\text{d,comp}}\right\}, T_{\text{E,o}}, H_{\text{chwp}}, \Delta T_{\text{C}}, T_{\text{C,i}}, T_{\text{sa}}, H_{\text{sa}}\right).\end{aligned} \tag{7.5}$$

7.2.2 Constraint Conditions

To ensure all parameters in the optimization process remain within a reasonable range, constraints on the boundaries of relevant parameters are necessary. Firstly, the number of operating units for each component type should be restricted to a nonzero vector, ensuring that at least one unit of each component is operating, as expressed in Eq. (7.6).

$$N_{\text{d, comp}} \neq \{0\}. \tag{7.6}$$

Secondly, the optimization parameters should be constrained within the feasible solution space, as shown in Eq. (7.7). In which, x_i is the ith optimization parameter. In this research, the dimension of optimization parameter d is 10, x_1–x_4 represent $N_{\text{d,ch}}$, $N_{\text{d, chwp}}$, $N_{\text{d, cwp}}$, and $N_{\text{d, ct}}$, respectively. And x_5–x_{10} represent $T_{\text{E,o}}$, H_{chwp}, ΔT_{C}, $T_{\text{C,i}}$, T_{sa}, and H_{sa}, respectively.

$$L_{\min} \le X^* \le L_{\max}, \quad X = \{x_i\}, \quad i = 1, 2, \ldots, d. \tag{7.7}$$

Additionally, the operating conditions for each unit of parallel components should be constrained within an acceptable range, ensuring that the allocation for each unit stays between the upper and lower limits, as expressed in Eq. (7.8).

$$S_{\min} \le S \le S_{\max}, \quad S \in \{Q_{\text{ch},i}, m_{\text{chw},j}, m_{\text{cw},m}, m_{\text{cw},n}, m_{\text{a},n}, m_{\text{sa},k}\}. \tag{7.8}$$

7.2.3 *Penalty Function*

To efficiently solve the above-constrained nonlinear optimization problem, a penalty function is often introduced into the original problem to convert it into an unconstrained optimization problem. The constraints for optimization parameters in Eqs. (7.6) and (7.7) will be reflected in the solution space of the optimization algorithm and will not be converted into penalty functions. However, the constraints for operating conditions in Eq. (7.8) will be converted into an unconstrained problem by constructing a penalty function. Thus, the constrained optimization problem expressed in Eq. (7.9) can be converted into an unconstrained optimization problem expressed in Eq. (7.10). Here, $P_i(x_i)$ is the penalty function, which can be expressed as Eq. (7.11); J_i is the penalty multipliers and should be set as a large positive number.

$$\min f(\text{X}), \quad \text{s.t.} \quad s_{i,\min} \le s_i(x_i) \le s_{i,\max}, \tag{7.9}$$

$$\min F(X) = \min\left[f(X) + \sum_{i=1}^{p} P_i(x_i)\right], \tag{7.10}$$

$$P_i(x_i) = \begin{cases} 0, & s_{i,\min} \le s_i(x_i) \le s_{i,\max} \\ J_i\left[s_i(x_i) - s_{i,\min}\right]^2, & s_i(x_i) \le s_{i,\min} \\ J_i\left[s_i(x_i) - s_{i,\max}\right]^2, & s_i(x_i) \ge s_{i,\max} \end{cases}. \tag{7.11}$$

7.3 Intelligent Optimization Algorithms

Compared to traditional solving methods, heuristic search algorithms offer significant advantages in tackling complex optimization problems with multiple dimensions, constraints, and nonlinearity. The Differential Evolution (DE) algorithm, a type of heuristic search algorithm, combines the strengths of the classical genetic algorithm (GA) while reducing algorithm complexity. It demonstrates robust global search capability and convergence speed. The DE algorithm has applications in various fields, including energy, intelligent control, network security, and economics, yielding commendable optimization results. The algorithm is straightforward in principle, featuring high search efficiency, making it highly suitable for addressing complex nonlinear optimization problems, such as the global optimization of HVAC water systems, and it possesses high online feasibility.

7.3.1 Exhaustive Search Method

Although the traditional Exhaustive Search (ES) method is practical and feasible for real systems, it encounters significant limitations when applied to complex air-conditioning water systems. As system complexity increases, the accuracy and efficiency of the algorithm decline, impacting the effectiveness of optimization control. The ES algorithm typically discretizes the feasible solution space and organizes it into a set of solutions. It then traverses all these solutions, compares the results, and identifies the optimal one. As a result, the ES algorithm can only provide non-inferior solutions, and its computational complexity grows exponentially with increased search accuracy. For instance, in an m-dimensional optimization problem, achieving a search accuracy of one-nth of the feasible solution interval requires n^m iterations to find the optimal solution at the current accuracy level. Due to these limitations, finding the optimal solution within the feasible space becomes challenging. In simpler systems, such as depicted in Fig. 7.5a, the ES algorithm remains practical and reliable. However, in more complex systems, like those in Fig. 7.5b, where optimization dimensions and search space expand, the ES algorithm consumes substantial resources, leading to poor efficiency and difficulties in ensuring result accuracy.

Unlike the Exhaustive Search (ES) algorithm, the heuristic search algorithm does not examine the entire feasible solution space. Instead, it assesses each state within the feasible space and directs the search toward the best fitness values until the optimal solution is located. This method greatly enhances both search efficiency and accuracy. Common heuristic search algorithms include genetic algorithms, ant

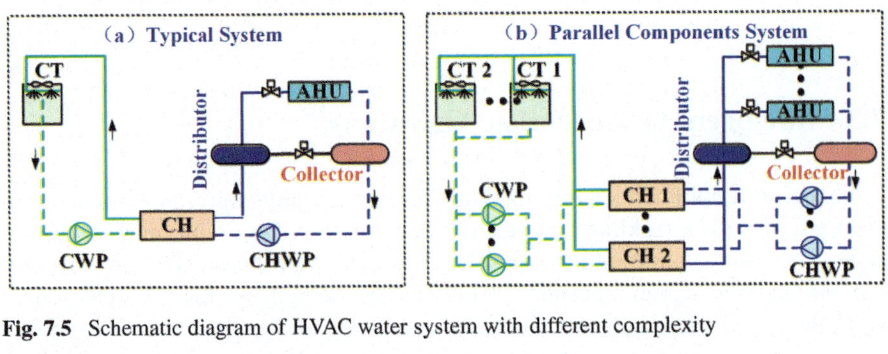

Fig. 7.5 Schematic diagram of HVAC water system with different complexity

colony algorithms, and particle swarm algorithms. These algorithms, inspired by natural biological processes, are extensively used in areas like artificial intelligence and distributed control.

Considering the complex and varied spatial structure of air-conditioning water systems, and the benefits of heuristic search algorithms, it is crucial to integrate the spatial characteristics of these systems with heuristic search methods. Researching global energy efficiency strategies using heuristic algorithms will help improve the overall operational performance of the system from a spatial perspective.

7.3.2 *Differential Evolution*

The Differential Evolution (DE) algorithm, introduced by R. Storn and K. Price in 1995, builds on the concept of population evolution found in genetic algorithms (GA). It maintains population diversity and fitness by simulating biological processes, such as mutation, crossover, and selection. Through multiple iterations, DE identifies the optimal individual within the population. As a heuristic optimization technique based on population differences, DE is well suited for finding global optimal solutions to multi-dimensional nonlinear optimization problems.

The solving process of the DE algorithm, illustrated in Fig. 7.6, involves four main steps: population initialization, mutation, crossover, and selection operations. Firstly, during population initialization, control parameters, such as population size (M), individual dimensions (n), feasible solution space ($L = [L_{\min}, L_{\max}]$), scaling factor (F), and crossover rate (cr), are initialized. Once parameter initialization is completed, the initial population $\{X_{i,0}\}$ can be generated, as represented by Eq. (7.12).

$$X_i^0 = \left(x_{i,1}^0, x_{i,2}^0, \ldots, x_{i,j}^0, \ldots, x_{i,n}^0\right), \quad i = 1, 2, \ldots M \tag{7.12}$$

The initialization value of the jth element ($x_{i,j}$) of the ith individual can be obtained according to Eq. (7.13). Here, $l = [l_{\min}, l_{\max}]$ represents the feasible solution space of the jth dimension of the individual, and rand(a, b) denotes a random floating-point number between a and b.

$$x_{i,j}^0 = l_{\min,j} + \text{rand}(0, 1) \cdot \left(l_{\max,j} - l_{\min,j}\right). \tag{7.13}$$

Next, the population undergoes mutation. Taking the mutation of the ith individual as an example, three random individuals ($r1$, $r2$, and $r3$) are selected from the population to perform the operation described in Eq. (7.14). Here, Hit represents the mutated individual i at generation t, and F is the scaling factor, typically ranging between [0, 2] and commonly set to 0.5.

$$H_i^t = X_{r1}^t + F \cdot \left(X_{r2}^t - X_{r3}^t\right). \tag{7.14}$$

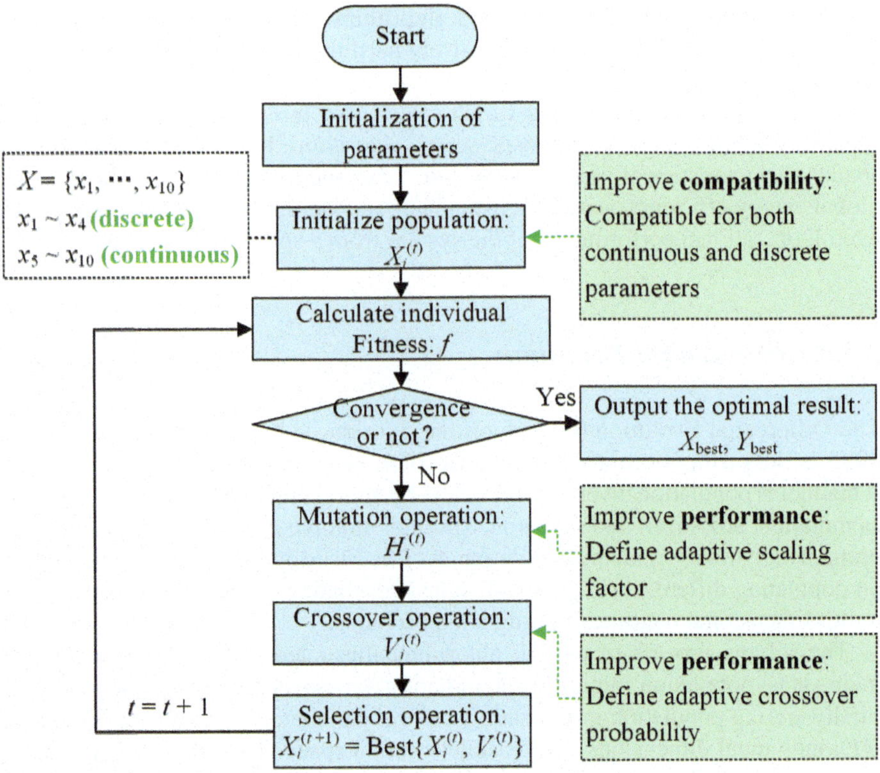

Fig. 7.6 Flow chart of DE algorithm and improvements (redrawn from Ref. [5], copyright 2022, with permission from ELSEVIER)

In the DE algorithm, mutation occurs within the parent population, directly selecting from both parents and offspring, and entirely discarding inferior individuals. Compared to the GA algorithm, DE inherently maintains higher population diversity and convergence speed, leading to more effective approximations when solving optimization problems.

$$v_{i,j}^t = \begin{cases} h_{i,j}^t, & \text{rand}(0,1) \le \text{cr} \\ x_{i,j}^t, & \text{otherwise} \end{cases}. \tag{7.15}$$

After mutation, crossover operations are performed between individuals and their mutated counterparts. This involves selecting, based on the crossover rate (cr), between the individual and the mutated individual for each dimension, as shown in Eq. (7.15). Here, $h_{i,j}$ and $v_{i,j}$, respectively, represent the jth element of the mutated individual H_i and the crossover individual V_i, with cr taking values between [0, 1].

Finally, selection is performed between the current individual and the crossover individual, with the fitter individual advancing to the next generation's computation, as indicated in Eq. (7.16). As iterations continue, the population progressively converges toward the optimal solution, ultimately yielding the best solution when the convergence criteria are satisfied.

$$X_i^{t+1} = \begin{cases} V_i^t, f\left(V_i^t\right) > f\left(X_i^t\right) \\ X_i^t, \text{ otherwise} \end{cases}. \tag{7.16}$$

7.3.3 Improvement for Integer Programming Problem

Based on the objective function outlined in Sect. 5.1.2, the global optimization problem for the HVAC water system is a mixed optimization problem. It involves both continuous variables, such as the setpoints of each control loop, and discrete variables, such as the operational counts (or *ON/OFF* states) of each piece of equipment. Since these counts must be integers, optimizing them involves integer programming. Traditional continuous Differential Evolution (DE) algorithms are not suitable for this type of problem. Therefore, it is necessary to adapt the DE algorithm to handle mixed optimization problems, allowing it to simultaneously address both continuous and discrete variables.

In optimizing the operating numbers (or *ON/OFF* states) for HVAC water systems, if the performance curves of similar equipment are identical, the *ON/OFF* states can be disregarded, focusing solely on optimizing the number of units. In this scenario, the optimization objective is the number of operating units ($N_{\mathrm{r,comp}}$) and the associated set values. However, when the performance curves of similar equipment differ, the *ON/OFF* states of each unit must be considered. Thus, the optimization results need to incorporate the ON/OFF state information for each piece of equipment. At this time, the optimization objective in the objective function is the on/off state of the equipment ($N_{\mathrm{d,comp}}$) and the various set values. Among them, $N_{\mathrm{r,comp}}$ is an integer variable with a range of $(0, N_{\mathrm{comp}}]$, and $N_{\mathrm{d,comp}}$ is a binary discrete vector composed of 0 and 1. Therefore, before optimization, $N_{\mathrm{d,comp}}$ needs to be converted to a decimal integer, and the converted equipment *ON/OFF* states have a range of N_{comp}. Hence, the optimization problems of the number of similar equipment and the *ON/OFF* state can be converted into integer programming problems.

To address the problem described, the DE algorithm is enhanced during the initialization and variation stages to handle both continuous and discrete states simultaneously. The main improvement involves applying value constraints to the integer dimensions of individuals. Specifically, if the *j*th dimensional element of individual *i* is an integer, rounding is applied to both the generated random individual and the varied individual for this dimension. For dimensions with real numbers, values are handled in the traditional manner. This adjustment ensures that the algorithm effectively manages mixed continuous and discrete optimization problems.

Equation (7.17) represents the value rule of the improved DE algorithm in the population initialization, where R and Z represent the real number set and the integer set, and round(a) represents the rounding operation of a according to the rounding principle. Similarly, the variation process can also be improved in this way, as shown in Eq. (7.18). The value rules of the improved initialization and variation processes correspond to the original Eqs. (7.13) and (7.14).

$$x_{i,j}^{0} = \begin{cases} l_{\min,j} + \text{rand}(0,1)\cdot\left(l_{\max,j} - l_{\min,j}\right), & x_{i,j} \in \{R\} \\ \text{round}\left[l_{\min,j} + \text{rand}(0,1)\cdot\left(l_{\max,j} - l_{\min,j}\right)\right], & x_{i,j} \in \{Z\} \end{cases}, \tag{7.17}$$

$$h_{i,j}^{t} = \begin{cases} x_{r1,j}^{t} + F\left(x_{r2,j}^{t} - x_{r3,j}^{t}\right), & x_{i,j} \in \{R\} \\ \text{round}\left[x_{r1,j}^{t} + F\left(x_{r2,j}^{t} - x_{r3,j}^{t}\right)\right], & x_{i,j} \in \{Z\} \end{cases}. \tag{7.18}$$

7.3.4 Construction of Adaptive Scaling Factor and Crossover Probability

The conventional Differential Evolution (DE) algorithm, while known for its rapid convergence, often struggles with local search capabilities. As iterations progress, the fixed scaling factor F and crossover rate cr used in the later stages can limit the accuracy of finding the global optimal solution, increasing the risk of converging to a local optimum. To address these issues and improve both the search capabilities and accuracy of the algorithm, we propose an enhancement: the Adaptive Hybrid Differential Evolution Algorithm (AH-DE). This improvement aims to better balance the algorithm's search effectiveness and accuracy.

First, in the mutation phase, the mutation operator is enhanced by expanding the conventional operator, which uses a single differential vector (DE/rand/1), to include two differential vectors (DE/rand to best/2), as shown in Eq. (7.19). This modification employs two differential vectors to improve both the convergence speed and global search capabilities of the algorithm. In the equation, $r1$, $r2$, and $r3$ represent three random individuals in the population, excluding individual i; b denotes the best individual in the current population.

$$h_{i,j}^{t} = \begin{cases} x_{r1,j}^{t} + F_i^t\left(x_{b,j}^{t} - x_{r1,j}^{t}\right) + G_i^t\left(x_{r2,j}^{t} - x_{r3,j}^{t}\right), & x_{i,j} \in \{R\} \\ \text{round}\left[x_{r1,j}^{t} + F_i^t\left(x_{b,j}^{t} - x_{r1,j}^{t}\right) + G_i^t\left(x_{r2,j}^{t} - x_{r3,j}^{t}\right)\right], & x_{i,j} \in \{Z\} \end{cases}. \tag{7.19}$$

F^t is the adaptive scaling factor, the size of which is related to the fitness value of the random individual, calculated by Eqs. (7.20) and (7.21). $F_{\min}$ and $F_{\max}$ are the boundaries of F^t, with values of 0.1 and 0.9 in this paper, f is the individual's fitness value, and $f_{\min}$ is the individual with the minimum fitness value among the three random individuals.

$$F_i^t = F_{\min} + (F_{\max} - F_{\min})\frac{f_b - f_{r1}}{f_b - f_{\min}}, \tag{7.20}$$

$$G_i^t = F_{\max} + F_{\min} - F_i^t. \tag{7.21}$$

The improved mutation operator has the following physical significance: When the fitness of the random individual $r1$ differs greatly from that of the best individual in the population, the adaptive scaling factor F_i is large, and G_i is small. This results in rapid mutation of individual i toward the best individual, enhancing convergence speed. Conversely, when $r1$'s fitness is similar to the best individual's fitness, F_i becomes small and G_i large. This slows the mutation of individual i toward the best individual, thereby improving local search capabilities and reducing the risk of premature convergence. Additionally, incorporating the second differential vector increases the randomness in the mutation process, which helps maintain population diversity.

In the crossover stage, the algorithm can be further enhanced by employing an adaptive crossover rate. Individuals are categorized into two groups based on their fitness relative to the population average. For individuals with fitness values above the average, the crossover rate is adjusted adaptively according to their fitness levels. For those with fitness values below the average, the crossover rate is set to its minimum value to preserve population diversity and maintain global search capabilities. This approach is detailed in Eq. (7.22), where f_{b} and f_{w} represent the fitness values of the best and worst individuals in the current population, $\overline{f}$ is the average fitness, and $\mathrm{cr}_{\min}$ and $\mathrm{cr}_{\max}$ denote the boundary values for the crossover rate, which are set to 0.1 and 0.6, respectively, in this study.

$$\mathrm{cr}_i = \begin{cases} \mathrm{cr}_{\min} + (\mathrm{cr}_{\max} - \mathrm{cr}_{\min})\frac{f_i - f_{\mathrm{w}}}{f_{\mathrm{b}} - f_{\mathrm{w}}}, & f_i > \overline{f} \\ \mathrm{cr}_{\min}, & f_i < \overline{f} \end{cases}. \tag{7.22}$$

The improved adaptive crossover rate can be understood as follows: When an individual's fitness is close to that of the best individual, a larger crossover rate is applied. This approach helps increase population diversity and accelerates convergence speed. Conversely, if an individual has a lower fitness value, a smaller crossover rate is used to enhance the global search capability and reduce the risk of premature convergence.

7.4 Comparison Research Preparation

7.4.1 Experimental System

The experiment in this chapter uses the air-conditioning system's topology shown in Fig. 7.7. Both optimization strategies aim for the best global system efficiency while meeting indoor cooling demands. Therefore, the objective function for the optimization strategy is to minimize the total electrical consumption, as shown in Eq. (7.23). To obtain acceptable optimization results, each optimization variable has a feasible domain, and the optimization process can only be carried out within this domain. The optimization process can be represented as Eq. (7.24) (* indicates the optimal solution), which derives the globally optimal combination solution through the energy efficiency game between each control loop. Considering the feasibility of online applications, this paper will use a relatively simple principle of the Exhaustive Search method to solve the global energy efficiency game problem.

$$f = \min P_{\text{sys}} = \min\left(P_{\text{ch}} + P_{\text{chwp}} + P_{\text{cwp}} + P_{\text{fan}}\right), \tag{7.23}$$

$$\left(T^*_{\text{C,i}}, \Delta T^*_{\text{C}}, \Delta T^*_{\text{E}}, T^*_{\text{sa}}, P^*_{\text{sa}}\right) = \arg\min P_{\text{sys}}\left(T_{\text{C,i}}, \Delta T_{\text{C}}, \Delta T_{\text{E}}, T_{\text{sa}}, P_{\text{sa}}\right). \tag{7.24}$$

The optimization results will be used as the setpoints for each control loop, including the condenser inlet water temperature ($T_{\text{C,i}}$), condenser inlet and outlet temperature difference (ΔT_{C}), evaporator inlet and outlet temperature difference (ΔT_{E}), supply air temperature (T_{sa}), and supply air static pressure (P_{sa}).

To ensure operational balance during the experiment, 12 typical days were selected for comparative experiments using SEDO and TDO strategies, resulting in 6 sets of experimental conditions. The observation period for each experiment group is

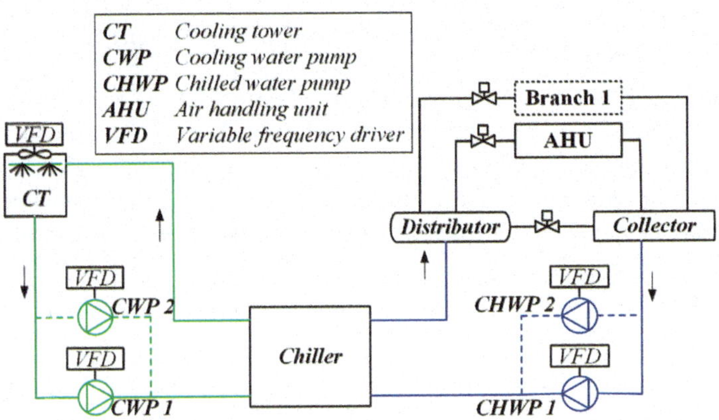

Fig. 7.7 Schematic of experimental system

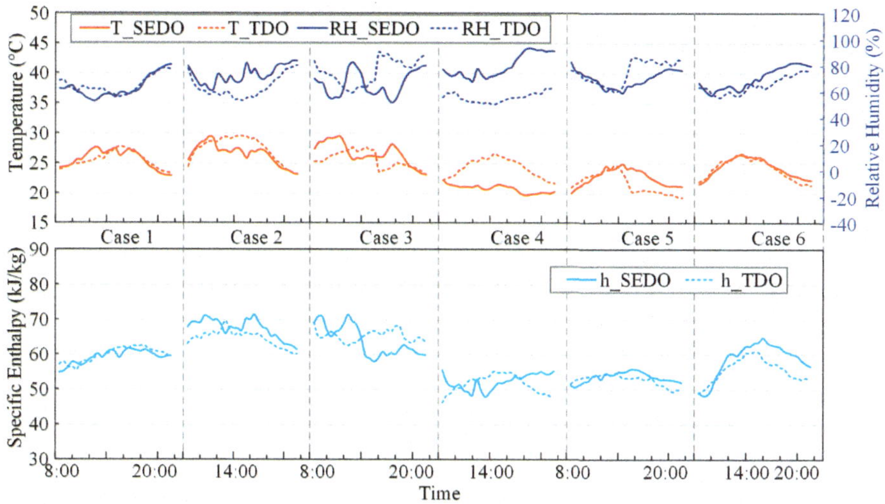

Fig. 7.8 Outdoor climate comparison of 6 groups of experiments (redrawn from Ref. [1], copyright 2022, with permission from ELSEVIER)

uniformly set from 8 a.m. to 9 p.m., totaling 13 h. Figure 7.8 shows the outdoor working condition comparison results for the 6 cases. It can be observed that the outdoor temperatures for both strategies range between 20 and 30 °C, and the enthalpy for the TDO strategy ranges from 46 to 70 kJ/kg, while for the SEDO strategy, it ranges from 48 to 72 kJ/kg. By comparing the outdoor condition parameters for both strategies, it can be seen that the outdoor load for the SEDO strategy is close to or slightly higher than that of the TDO strategy, which can be approximately considered to meet the outdoor condition balance condition. The comparative experimental results are informative. By comparing the enthalpy of the outdoor air, it is slightly higher than that of the TDO strategy. During the experiment, there is a regular schedule for people in the room, and there are no large-scale personnel movements, ensuring the balance of indoor conditions. The optimization period for the TDO strategy is 15 min, which is relatively short in practical applications [3] and can meet the requirements of this comparative experiment.

7.4.2 Evaluation Index for Control Effect

When evaluating the decoupling effect, there is currently no unified evaluation metric. Most studies tend to compare the disturbance of parameters before and after decoupling to judge its effectiveness. This is because coupling characteristics can exacerbate the disturbance of system operating parameters, causing certain parameters to deviate from their original stable state. Therefore, the effectiveness of decoupling can be evaluated by observing the disturbance levels of various parameters during the operating period. The smaller the disturbance level, the better the decoupling effect. Following this principle, this paper defines three indicators to quantify the decoupling effect: total steady-state duration (t_{sum}), steady-state continuity (t_{con}), and parameter dispersion (d_α).

Referring to the definition of steady state in control theory, the steady-state interval is defined as follows: for a given time point a_i, there exists a time point b_i, such that Eq. (7.25) holds for any t in the interval $[a_i, b_i]$, and does not hold for $t > b_i$. This interval $[a_i, b_i]$ is referred to as the steady-state interval.

$$\left| y(t) - \frac{1}{b_i - a_i} \cdot \sum_{t=a_i}^{b_i} y(t) \right| < \varepsilon, \quad t \in [a_i, b_i], \quad i \in [0, n], \quad t \geq t_{bou}. \tag{7.25}$$

Here, $y(t)$ represents the measured value of parameter t at time t, n is the total number of steady states observed, and ε and t_{bou} represent the acceptable range of $y(t)$ and the lower limit value of t.

Let the steady-state duration of the parameter in the interval $[a_i, b_i]$ be $t_i = b_i - a_i$. Then, t_{sum} is the total sum of all t_i observed during the observation period, as shown in Eq. (7.26), and t_{con} is the ratio of the average value of all t_i observed during the observation period to the observation period, as shown in Eq. (7.27). The calculation of t_{sum} and t_{con} is illustrated in Fig. 7.9.

$$t_{sum} = \sum_i t_i, \tag{7.26}$$

$$t_{con} = \frac{t_{sum}}{n t_{obs}} \times 100\%. \tag{7.27}$$

The parameter dispersion d_α evaluates the decoupling effect from the overall perspective of the observation period. Referring to the definition of the coefficient

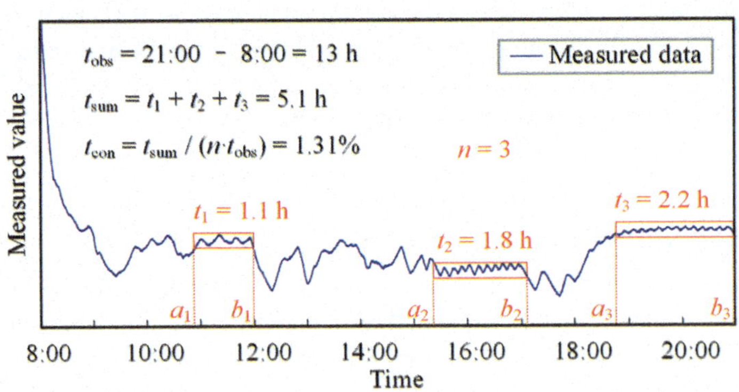

Fig. 7.9 Schematic diagram of steady-state period and continuity (redrawn from Ref. [1], copyright 2022, with permission from ELSEVIER)

of variation, the dispersion can be defined as the average relative deviation of all measured parameter values from the mean value during the observation period, as shown in Eq. (7.28):

$$d_\alpha = \frac{1}{\overline{y}} \cdot \sum_{i=1}^{m} \frac{|y_i - \overline{y}|}{m}. \tag{7.28}$$

Here, n is the total number of parameter measurements during the observation period, y_i represents the ith measurement value, and $\overline{y}$ represents the parameter mean.

Based on the above definitions, the effectiveness of parameter decoupling under the current strategy can be quantified using these three evaluation indicators. Among them, the total steady-state duration (t_{sum}) can reflect the total duration that parameters maintain a steady state during the observation period. A larger t_{sum} indicates a longer duration of steady state, suggesting that the parameters experience less fluctuation during the optimization control period. t_{con} can reflect the fragmentation of steady states, with a larger t_{con} indicating a longer average steady-state duration and better continuity of steady states. d_α reflects the overall fluctuation of parameters, with a smaller d_α indicating less overall fluctuation during the observation period, and conversely, more significant overall fluctuation and poorer overall stability of parameters.

According to the evaluation method of control effect in the reference control theory, the control effect of the optimization strategy can be quantified by using the transient time (t_s), overshoot (o_s), and steady-state error (s_d), as shown in the schematic diagram of the three evaluation indicators in Fig. 7.10. Among them, t_s can reflect the response speed of the system to reach a new steady state after being disturbed; o_s represents the ratio of the maximum deviation ($y_{\max} - y_{\text{set}}$) during the control process to the steady-state value (y_{set}), which can be expressed as Eq. (7.29); s_d represents the acceptable fluctuation range after the parameters reach a steady state.

$$o_s = \frac{y_{\max} - y_\infty}{y_\infty} \times 100\%. \tag{7.29}$$

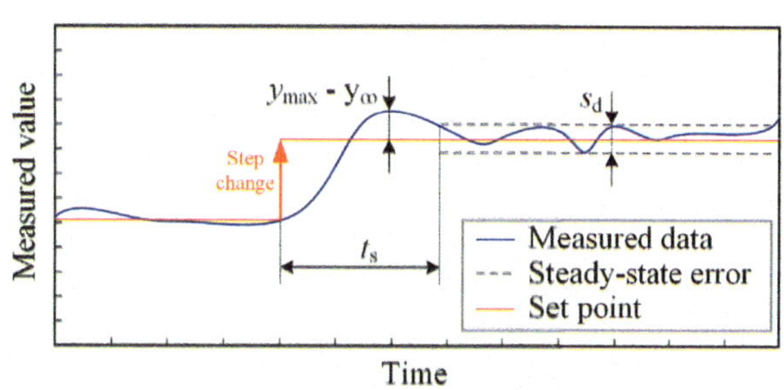

Fig. 7.10 Schematic diagram of evaluation index of control effect (redrawn from Ref. [1], copyright 2022, with permission from ELSEVIER)

In addition to the above three indicators, the introduction of the degree of discreteness (d_β) is also used to evaluate the overall control effect during the observation period. Its expression is similar to d_α, but d_α is the degree of discreteness for the mean, while d_β is the degree of discreteness for the setpoint, as shown in Eq. (7.30).

$$d_\beta = \frac{1}{m} \cdot \sum_{i=1}^{m} \frac{\left| y_i - y_{i,\text{set}} \right|}{y_{i,\text{set}}}. \tag{7.30}$$

Here, y_i is the parameter measurement value, $y_{\text{set},i}$ is the corresponding setpoint, and n is the total number of parameter measurements during the observation period. The smaller the d_β value, the better the overall control effect of the optimization strategy during the observation period.

7.4.3 Selection of Event Space

In Sect. 4.1.3 of the SEDO strategy, various commonly used methods for establishing the event space have been briefly outlined. Among them, the random forest feature selection method is highly feasible in practical applications and can be used to determine the state parameters required to construct the event space. Considering the online feasibility of event recognition, state parameters will be selected from parameters that are easily obtained directly, such as temperature and pressure. As mentioned in Sect. 4.1.3, the event space should be able to quickly capture changes in operating conditions and be closely related to the system's energy efficiency. Therefore, the state parameters of the event space can be determined based on the feature importance of the System Coefficient of Performance (SCOP). The higher the feature importance of a parameter to the SCOP, the greater its value as a state parameter for the event space.

The feature importance of the random forest algorithm can be calculated using various methods, with each method having different emphases. Typically, the results of multiple methods are combined for comprehensive evaluation. This paper will use three methods for comprehensive evaluation: Gini importance, permutation-based importance, and SHAP value-based importance. Gini importance selects features based on decision trees, providing fast computation speed, but its drawback is that the results tend to favor numerical and categorical features with higher cardinality, leading to inaccurate results. Permutation-based importance can overcome the drawback of calculating default feature importance based on average impurity reduction, but it has a higher computational cost. SHAP value-based importance uses Shapley values from cooperative game theory to estimate the contribution of each feature to the prediction results, providing richer information but also incurring higher computational costs.

In this analysis, four groups (a total of 10) of representative parameters related to the overall system operation characteristics will be selected for feature importance analysis, including 2 outdoor working condition parameters (outdoor air temperature T_{out} and relative humidity RH_{out}), 3 cooling water-side parameters (inlet water temperature $T_{C,i}$, outlet water temperature $T_{C,o}$, and temperature difference ΔT_C), 3 chilled water-side parameters (inlet water temperature $T_{E,i}$, outlet water temperature $T_{E,o}$, and temperature difference ΔT_E), and 2 air-side parameters (supply air temperature T_{sa} and return air temperature T_{ra}). These parameters can all be directly obtained in actual systems, ensuring the online feasibility of the event recognition process [6].

Figure 7.11 shows the statistical results of the importance of various parameters to the SCOP feature in historical data (June–October). The data was collected at 5-min intervals and includes a total of 9360 data points. Subfigures (a), (b), and (c) represent the results of feature importance calculated using three different methods, with the vertical axis showing the parameters arranged in descending order of importance and the horizontal axis indicating the degree of influence of each parameter on the output (SCOP). Subfigure (d) provides detailed calculation results based on the importance of SHAP values, showing the distribution of parameters under different SHAP results. The larger the absolute value of the SHAP value, the greater the correlation between the parameter and the output. Based on these statistical results, if only the most relevant parameter for each representative group is retained, the state parameters of the event space can be directly determined, as indicated by the red box in the figure. Therefore, these four state parameters can be used to describe the event space of the SEDO strategy, namely T_{out}, ΔT_C, $T_{E,i}$, and T_{ra}.

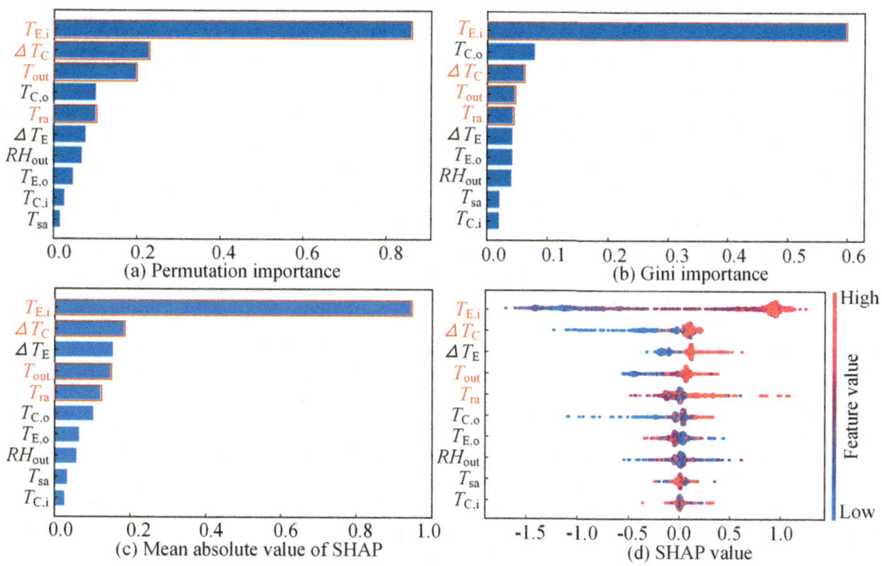

Fig. 7.11 Statistical results of importance of parameter features (redrawn from Ref. [1], copyright 2022, with permission from ELSEVIER)

Table 7.1 Event space and trigger conditions

Event	Conditions	Threshold
T_{out} significantly change	$\{M_s = ON\&\vert T_{out}^{prev} - T_{out}^{curr}\vert \geq \zeta T_{out}\}$	$\zeta T_{out} = 0.6$ °C
ΔT_C significantly change	$\{M_s = ON\&\vert \Delta T_C^{prev} - \Delta T_C^{curr}\vert \geq \zeta_{\Delta T_C}\}$	$\zeta_{\Delta T_C} = 0.5$ °C
$T_{E,i}$ significantly change	$\{M_s = ON\&\vert T_{E,i}^{prev} - T_{E,i}^{curr}\vert \geq \zeta T_{E,i}\}$	$\zeta T_{E,i} = 0.4$ °C
T_{ra} significantly change	$\{M_s = ON\&\vert T_{ra}^{prev} - T_{ra}^{curr}\vert \geq \zeta T_{ra}\}$	$\zeta T_{ra} = 0.9$ °C

This paper constructs the event space using the most commonly used definition method, defining an event as "significant changes in state parameters", with the determination criteria represented in the form of Eq. (7.2). The final determined event space is shown in Table 7.1. Here, Ms represents the state of the preparatory mode in the SEDO strategy, and ζ is the threshold of the determination criteria, calculated using data mining methods provided in Ref. [4]. This method continuously searches for the optimal solution in historical data by setting a reward mechanism until the optimal reward is obtained, thereby determining the best threshold for each event.

7.5 Performance Analysis

7.5.1 Optimization Timing

For the TDO strategy, since it has a periodic triggering mechanism, the number of optimizations in the 6 sets of comparative experiments is fixed. As for the SEDO strategy, optimization only occurs when the triggering conditions are met, and the number of optimizations depends on the actual operating conditions.

The statistical results of the optimization times for the two strategies in the 6 sets of experiments are shown in Table 7.2. From the table, it can be seen that the optimization times for the SEDO strategy are much higher than those for the TDO strategy, which is inconsistent with the research results on event-driven strategies in Refs. [4, 7]. The reasons for this difference can be attributed to differences in system operational characteristics and event triggering mechanisms.

On the one hand, the experimental system's chilled water unit consists of multiple compressor water-cooled units. During the experiment, to meet the actual cooling load requirements, there are often situations where the number of compressor units

Table 7.2 Statistics result of accumulative optimization times

	Test 1	Test 2	Test 3	Test 4	Test 5	Test 6
TDO (times)	52	52	52	52	52	52
SEDO (times)	53	116	158	81	83	76

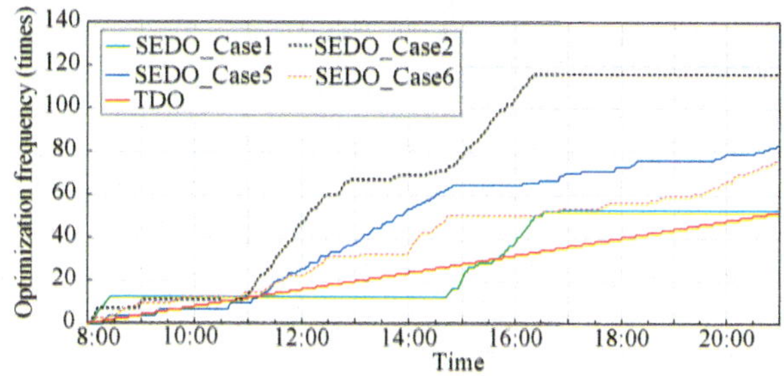

Fig. 7.12 Comparison results of cumulative optimization times (redrawn from Ref. [1], copyright 2022, with permission from ELSEVIER)

switches, leading to a higher event triggering frequency as defined in Table 7.1. On the other hand, the triggering interval of the proposed SEDO strategy is related to the parameter time-delay response time and the duration of the standby state, without the need to separately set a minimum triggering interval. In contrast, the EDO strategy in the literature usually sets a minimum triggering interval (such as 5 min [4] or 10 min [7]) based on operational experience to avoid frequent event triggering. This is also one of the reasons for the differences in research results.

Figure 7.12 shows the comparison of hourly cumulative optimization frequencies in some experiments, demonstrating the differences in optimization frequency between the SEDO and TDO strategies. From the figure, it can be observed that the optimization frequency of the TDO strategy increases at a fixed step length from the beginning to the end of the experiment, while the value for the SEDO strategy increases at a variable step length. Although the overall optimization frequency of the SEDO strategy is relatively frequent, its triggering periods are concentrated, mainly distributed in three periods of intense load changes: the initial stage of the experiment, from 11 a.m. to 2 p.m., and from 3 p.m. to 5 p.m., corresponding to the initial stage of system operation, the rise in outdoor load, and the decline in outdoor load, respectively. This indicates that the SEDO strategy can perform timely and necessary optimization control based on load changes.

Table 7.3 presents the statistical results of the damper action times at the end of the variable air volume for the two strategies. A record is made each time the damper's control signal changes. It can be seen that during the experiment, on average, the SEDO strategy can save 63 (17.1%) damper action times compared to the TDO strategy, indicating that the SEDO strategy has better economic benefits. Although the optimization frequency of the SEDO strategy is relatively high, the number of damper actions is relatively small. This is because the SEDO strategy can trigger optimization selectively through event recognition, reducing unnecessary control actions. Moreover, the overall system operation fluctuation caused by optimization actions is relatively small, allowing the system to maintain a relatively stable operating state. When adjustments are needed, only minor adjustments are required to meet the demand.

Table 7.3 Statistics on number of damper actions (data from Ref. [1], copyright 2022, with permission from ELSEVIER)

Area	Method	Test 1	Test 2	Test 3	Test 4	Test 5	Test 6	Average
Room 1	SEDO	158	68	66	50	66	98	84
	TDO	154	94	132	66	56	90	99
Room 2	SEDO	88	78	102	102	96	88	92
	TDO	122	110	98	56	176	110	112
Room 3	SEDO	154	86	76	106	194	154	128
	TDO	166	124	152	102	164	236	157
All rooms	SEDO	400	232	244	258	356	340	305
	TDO	442	328	382	224	396	436	368

7.5.2 Decoupling Effect

For the decoupling effect of the two optimization strategies, the evaluation mainly focuses on the decoupling of the system's quality parameters and is quantitatively analyzed according to the evaluation indicators. The evaluation objects include the water-side temperature and flow, and the air-side temperature and flow.

Tables 7.4 and 7.5, respectively, show the statistical results of the decoupling effect on the water side and the air side in the six experiments. From the tables, it can be seen that the t_{sum} and t_{con} of the SEDO strategy are generally better than the TDO strategy. In all six experiments, the average t_{sum} values for supply water temperature, water flow, supply air temperature, and supply air volume are all higher by 1.9 h, 2.6 h, 0.7 h, and 1.7 h, respectively, compared to the TDO strategy. This indicates that the SEDO strategy can achieve the decoupling optimization of quality parameters while considering the time-delay response characteristics of the working fluid, thereby extending the duration of steady state. Similarly, comparing the average t_{con} values, it is evident that the SEDO strategy has a longer average duration for all steady-state periods compared to the TDO strategy. This means that the SEDO strategy can maintain stability for a longer period by reducing unnecessary optimization operations, allowing the system to remain in a steady state for a longer time.

By comparing the t_{sum} values of quality parameters and quantity parameters under the same strategy, it can be seen that the t_{sum} and t_{con} values of quantity parameters are both larger than those of quality parameters. This can be explained by the difference in response speed between the two parameters. Because quantity parameters (such as pressure and flow rate) have a faster response speed, they will reach a steady state faster than quality parameters (such as temperature) under the same external disturbance and have a higher steady-state duration and continuity.

Figures 7.13 and 7.14 show the average dispersion (d_{α}) of the water-side and air-side parameters. From the figures, it can be visually observed that the d_{α} values of each parameter under the SEDO strategy are lower than those under the TDO strategy (except for test 2), indicating that the overall fluctuation of system operating

Table 7.4 Statistical results of decoupling effect of parameters on water side (data from Ref. [1], copyright 2022, with permission from ELSEVIER)

Parameter	Index	Method	Test 1	Test 2	Test 3	Test 4	Test 5	Test 6	Average
Water supply temperature	t_{sum} (h)	SEDO	4.8	8.6	9.1	9.3	4.5	5.3	6.9
		TDO	4.5	7.0	4.6	6.0	5.0	3.0	5.0
	t_{con}	SEDO (%)	18.5	33.1	35.0	23.8	17.3	20.4	24.7
		TDO (%)	17.3	26.9	17.7	23.1	19.2	11.5	19.3
Water flow	t_{sum} (h)	SEDO	11.4	7.4	10.8	12.8	12.6	9.6	10.8
		TDO	8.2	4.8	6.6	12.4	8.0	9.4	8.2
	t_{con}	SEDO (%)	21.9	19.0	27.7	98.5	96.9	24.6	48.1
		TDO (%)	21.0	9.2	16.9	95.4	30.8	36.2	34.9

Table 7.5 Statistical results of decoupling effect of parameters on air side (data from Ref. [1], copyright 2022, with permission from ELSEVIER)

Parameter	Index	Method	Test 1	Test 2	Test 3	Test 4	Test 5	Test 6	Average
Air supply temperature	t_{sum} (h)	SEDO	4.8	9.4	7.5	5.1	2.6	4.0	5.6
		TDO	4.0	6.7	6.4	5.4	3.8	2.8	4.9
	t_{con}	SEDO (%)	18.5	72.3	57.7	13.1	10.0	15.4	31.2
		TDO (%)	30.8	25.8	49.2	10.4	14.6	10.8	23.6
Air supply volume	t_{sum} (h)	SEDO	4.8	6.0	8.0	10.6	4.9	5.6	6.7
		TDO	5.0	6.0	5.0	5.6	4.0	4.2	5.0
	t_{con}	SEDO (%)	18.5	46.2	61.5	40.8	18.8	21.5	34.6
		TDO (%)	19.2	46.2	38.5	21.5	15.4	16.2	26.2

parameters is smaller under the SEDO strategy during the experiment period. In addition, there is no significant difference between the two strategies in the results of test 2, and the overall fluctuation of air-side parameters under the SEDO strategy is even more intense than that under the TDO strategy, which may be caused by the significant fluctuation of the outdoor enthalpy value under the SEDO strategy during the experiment period.

7.5.3 Control Effect

When evaluating the control effects of the two optimization strategies, the main focus is on the control loop's ability to track the setpoint. Quantitative analysis is carried out based on the evaluation index of control effect. The setpoint parameters include supply water temperature, supply air temperature, supply air static pressure, and indoor temperature. Tables 7.6 and 7.7 present the statistical results of the control effects on the water side and air side in six sets of experiments, respectively. In

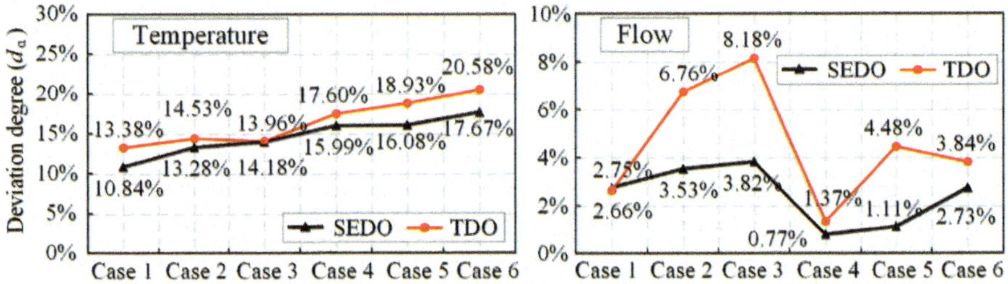

Fig. 7.13 Comparison results of decoupling effect (d_α) of parameters on water side (redrawn from Ref. [1], copyright 2022, with permission from ELSEVIER)

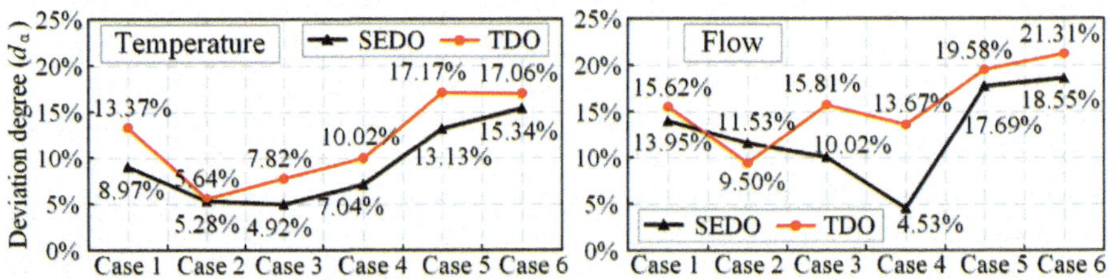

Fig. 7.14 Comparison results of decoupling effect (d_α) of parameters on air side (redrawn from Ref. [1], copyright 2022, with permission from ELSEVIER)

the tables, the t_s value represents the average transition time when changing the setpoint during the experimental observation period, while the o_s value represents the maximum overshoot. The s_d value remains consistent within the same set of experiments. It can be observed that the s_d values for supply air temperature and supply air static pressure are uniformly $\pm$ 5% and $\pm$ 10%, respectively. However, the s_d value for supply water temperature varies across different experiments. This is due to the unique control method for the outlet water temperature of the chiller unit, which is controlled by the start/stop strategy of the compressor. When different numbers of compressors are activated, the fluctuation range of the outlet temperature will vary. Therefore, this paper sets three groups of s_d values for supply water temperature in different experiments ($\pm$ 5%, $\pm$ 15%, and $\pm$ 28%).

From the statistical results, it can be seen that there is no significant difference in the maximum overshoot (o_s) of the three control variables between the two strategies. This is because both strategies employ PI control algorithms to track the setpoint for each basic control loop during the experiment, resulting in similar maximum overshoot values. For the three control variables, with setpoints of 9 °C, 16 °C, and 80 Pa, for example, the maximum overshoot values under both strategies are around 2.7 °C, 3.2 °C, and 24 Pa, which fall within the acceptable range for the experiment.

By comparing the t_s values, it can be seen that the average transition time under the SEDO strategy is lower than that under the TDO strategy. The average transition times for the three control variables are 10.8 min, 8.3 min, and 2.8 min under the SEDO strategy, indicating that the control action under the SEDO strategy is timelier and

Table 7.6 Statistical results of control effect on water supply temperature (data from Ref. [1], copyright 2022, with permission from ELSEVIER)

Index	Method	Test 1	Test 2	Test 3	Test 4	Test 5	Test 6	Average
t_s (min)	SEDO	12.2	11.5	14.4	17.4	21.5	12.2	14.9
	TDO	20.2	27.0	31.2	25.9	33.5	16.6	25.7
o_s	SEDO (%)	22.0	16.7	27.8	41.4	27.5	27.8	27.2
	TDO (%)	15.0	18.9	30.0	40.0	34.3	31.3	28.2
s_d	SEDO (%)	± 5	± 15	± 15	± 28	± 15	± 15	\
	TDO (%)	± 5	± 15	± 15	± 28	± 15	± 15	\

Table 7.7 Statistical results of control effect on air supply temperature and static pressure (data from Ref. [1], copyright 2022, with permission from ELSEVIER)

Control variable	Index	Method	Test 1	Test 2	Test 3	Test 4	Test 5	Test 6	Average
Air supply temperature	t_s (min)	SEDO	7.7	8.6	13.4	13.9	20.2	14.8	13.1
		TDO	9.5	19.1	31.7	18.0	28.8	21.4	21.4
	o_s	SEDO (%)	10.7	12.5	6.9	7.2	34.4	36.7	18.0
		TDO (%)	18.1	9.4	5.0	10.8	40.0	36.7	20.0
	s_d	SEDO (%)	± 5	± 5	± 5	± 5	± 5	± 5	\
		TDO (%)	± 5	± 5	± 5	± 5	± 5	± 5	\
Static pressure	t_s (min)	SEDO	7.2	10.6	7.2	8.4	6.7	7.2	7.9
		TDO	8.6	11.5	9.6	11.9	10.6	11.8	10.7
	o_s	SEDO (%)	37.5	37.5	33.3	21.6	14.0	19.2	27.2
		TDO (%)	37.5	37.5	20.0	26.0	27.3	22.2	28.4
	s_d	SEDO (%)	± 10	± 10	± 10	± 10	± 10	± 10	\
		TDO (%)	± 10	± 10	± 10	± 10	± 10	± 10	\

more accurate, avoiding significant deviations from the control target and allowing the system to be adjusted to the set state in a relatively short time. In addition, under the same control strategy, the t_s for water temperature is the longest, while the t_s for static pressure is the shortest, consistent with the differences in response speed of the control variables; that is, the pressure response speed is faster, requiring a relatively short transition time, while the water temperature response speed is slower, requiring a relatively long transition time.

The average dispersion (d_β) of the three control variables (supply water temperature, supply air temperature, and supply air static pressure) from the setpoint during the experimental observation period is shown in Figs. 7.15 and 7.16. From the figures, it can be seen that the d_β values under the SEDO strategy are generally lower than those under the TDO strategy. The average d_β values for the three control variables under the SEDO strategy are 15.6%, 11.7%, and 17.4%, while those under the TDO strategy are 17.9%, 14.6%, and 20.7%, respectively. Similarly, taking setpoints of

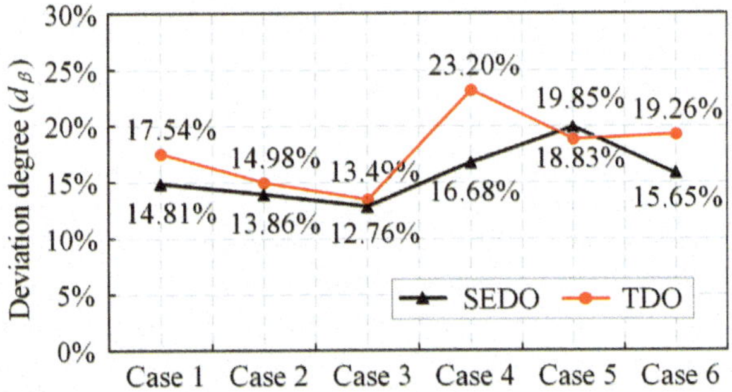

Fig. 7.15 Comparison results of control effect (d_β) on water supply temperature (redrawn from Ref. [1], copyright 2022, with permission from ELSEVIER)

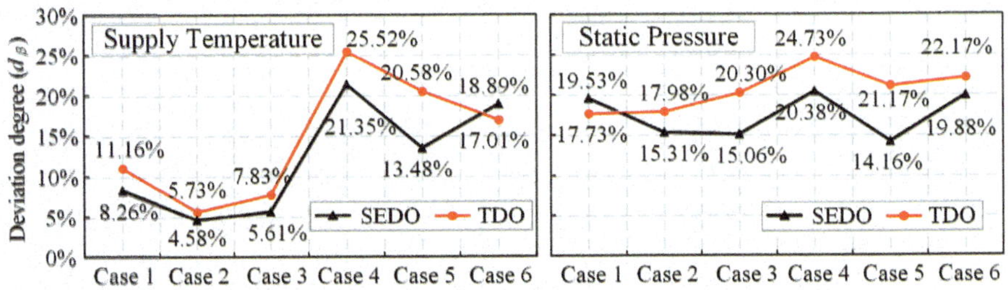

Fig. 7.16 Comparison results of control effect (d_β) on air temperature and static pressure (redrawn from Ref. [1], copyright 2022, with permission from ELSEVIER)

9 °C, 16 °C, and 80 Pa as examples, the deviations of the measured values of the control variables from the setpoints under the SEDO strategy are 1.4 °C, 1.9 °C, and 13.9 Pa, while those under the TDO strategy are 1.6 °C, 2.3 °C, and 16.6 Pa. From a global perspective, the control algorithm under the SEDO strategy performs better in tracking the control target.

In Table 7.8 and Fig. 7.17, the control effect of indoor temperature is assessed with a standard deviation (s_d) uniformly set at $\pm 2\%$. Figure 7.17 depicts the scenario with an indoor temperature setpoint of 26 °C. Under the SEDO strategy, the average deviation of indoor temperatures from the setpoint is 0.4 °C, 0.5 °C, and 0.5 °C for the three rooms. In contrast, the TDO strategy yields average deviations of 0.7 °C, 0.7 °C, and 0.6 °C for the same rooms. From the perspective of indoor comfort, both strategies exhibit control effects within an acceptable range. However, in terms of control accuracy, the SEDO strategy demonstrates a superior ability to precisely track the temperature setpoint compared to the TDO strategy. Moreover, the control process of indoor temperature under the SEDO strategy reveals a shorter transition time and smaller deviation from the setpoint, indicating an overall better control effect for the system under the SEDO strategy as opposed to the TDO strategy.

Table 7.8 Statistical results of control effect on indoor temperature (data from Ref. [1], copyright 2022, with permission from ELSEVIER)

Area	Index	Method	Test 1	Test 2	Test 3	Test 4	Test 5	Test 6	Average
Room 1	t_s (min)	SEDO	8.6	8.3	9.0	9.8	14.6	22.1	12.1
		TDO	9.7	12.2	14.7	16.8	19.3	27.6	16.7
	o_s	SEDO (%)	2.3	2.4	2.7	4.7	3.1	2.7	3.0
		TDO (%)	2.3	3.1	2.3	2.8	3.5	3.3	2.9
	s_d	SEDO (%)	± 2	± 2	± 2	± 2	± 2	± 2	\
		TDO (%)	± 2	± 2	± 2	± 2	± 2	± 2	\
Room 2	t_s (min)	SEDO	8.3	11.5	12.2	13.4	7.9	12.9	11.1
		TDO	7.6	19.8	22.1	15.1	8.4	12.2	14.2
	o_s	SEDO (%)	4.4	2.2	2.6	2.3	4.2	2.3	3.0
		TDO (%)	4.4	2.6	2.2	3.5	4.6	3.5	3.5
	s_d	SEDO (%)	± 2	± 2	± 2	± 2	± 2	± 2	\
		TDO (%)	± 2	± 2	± 2	± 2	± 2	± 2	\
Room 3	t_s (min)	SEDO	11.3	11.5	13.8	11.6	14.6	17.0	13.3
		TDO	14.4	17.4	19.8	17.1	13.8	21.1	17.3
	o_s	SEDO (%)	3.2	2.4	4.8	2.0	4.6	4.6	3.6
		TDO (%)	3.2	2.1	4.3	3.6	4.4	4.7	3.7
	s_d	SEDO (%)	± 2	± 2	± 2	± 2	± 2	± 2	\
		TDO (%)	± 2	± 2	± 2	± 2	± 2	± 2	\

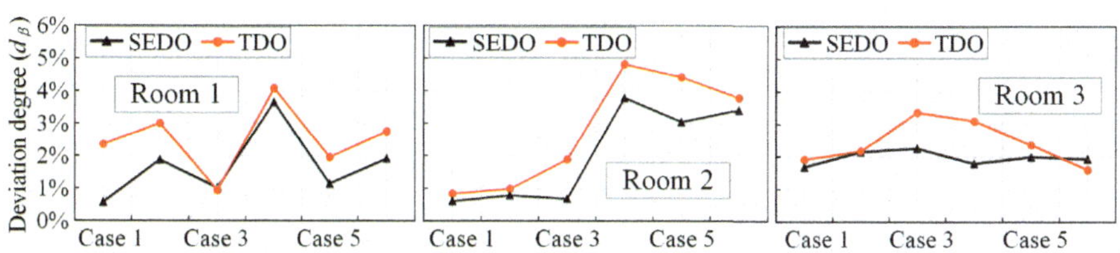

Fig. 7.17 Comparison results of control effect (d_β) on indoor temperature (redrawn from Ref. [1], copyright 2022, with permission from ELSEVIER)

7.5.4 Energy Saving Potential

To assess the energy-saving potential of the two optimization strategies, the evaluation primarily focuses on the statistical power consumption of the system equipment. In Fig. 7.18, the total power consumption of the system and the power consumption of the equipment in six comparative experiments were analyzed. The percentage in the figure represents the relative energy savings achieved by the SEDO strategy compared to the TDO strategy.

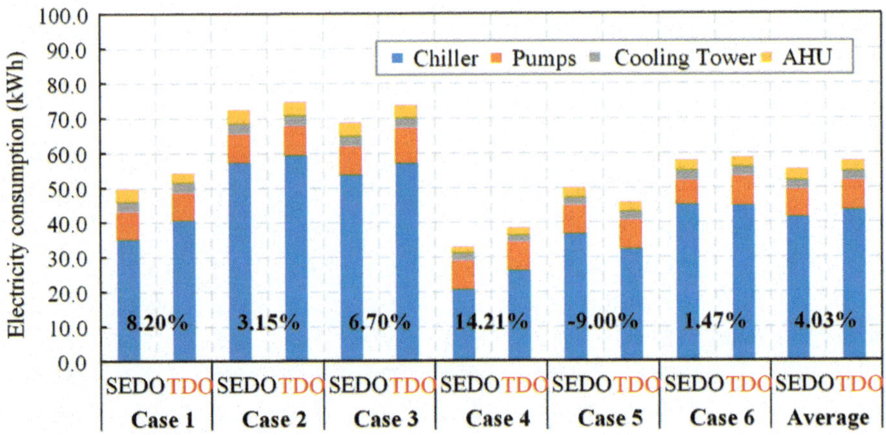

Fig. 7.18 Comparison results of system power consumption (redrawn from Ref. [1], copyright 2022, with permission from ELSEVIER)

Analyzing the outdoor weather conditions depicted in Fig. 7.8, it becomes apparent that the SEDO strategy generally exhibits high energy-saving potential, particularly in situations with elevated outdoor loads (such as test 1, test 2, and test 3). Notably, even when the outdoor load under the SEDO strategy in test 2 surpasses that under the TDO strategy, its power consumption is still 3.15% lower than that of the TDO strategy. Comparing the results from the six experiments, it is evident that the SEDO strategy proposed in this paper can potentially save approximately 4.0% of electricity consumption per day compared to the TDO strategy.

7.6 Summary

Traditional fixed-period optimization strategies often lead to "untimely" or "unnecessary" optimization actions due to ignoring the temporal characteristics of the system, resulting in less-than-ideal decoupling control effects during actual system operation. In response to this issue, this chapter firstly proposes a variable-period optimization strategy based on sequential event-driven method. Through comparative experimental studies with traditional time-driven optimization strategies, the SEDO strategy is validated to have advantages in decoupling effects, control effects, and energy-saving potential. And then, this chapter focuses on the efficiency and accuracy issues in solving the global energy efficiency game problem of complex HVAC water systems using traditional optimization methods. By combining the spatial characteristics of the system, a thorough investigation of the global energy efficiency game problem and optimization methods has been conducted, clarifying the correlation between the operating parameters of each device, the spatially related operating parameters of devices, and their own energy consumption. Based on the solution idea of heuristic search algorithms, an Adaptive Hybrid Differential Evolution Algorithm is proposed. Through simulation and experimental studies, the reliability of the algorithm has been verified in terms of computational efficiency, solution accuracy, and energy-saving potential, effectively addressing the spatial coupling issues between devices and control loops. The content of this chapter provides important theoretical guidance and practical reference value for studying the global energy efficiency game methods of central HVAC water systems and improving the actual global operational energy efficiency of HVAC water systems.

Based on the research content of this chapter, the following conclusions are drawn:

(1) Compared to traditional TDO strategies, the SEDO strategy demonstrates better decoupling and control effects, with overall fluctuations of quality-mass parameters during system operation being smaller, making it easier to achieve control objectives. In terms of energy-saving potential, the SEDO strategy offers a 4.0% reduction in electricity consumption compared to TDO strategies, providing better economic benefits.
(2) The AH-DE algorithm proposed in this paper can adaptively adjust the crossover rate and scaling factor based on the changes in individual fitness, effectively improving the algorithm's search accuracy and avoiding falling into local optima. Meanwhile, the mutation operator of the algorithm comprehensively considers factors, such as population diversity and convergence speed. While enhancing search accuracy, it can also maintain good computational efficiency.
(3) Compared to the fixed setpoint control method that does not consider system spatial characteristics, the AH-DE algorithm can save up to 16.1% of electricity consumption under typical daily operating conditions and 17.2% of electricity consumption during the entire cooling season.
(4) Compared to traditional GA and DE algorithms, the performance of the AH-DE algorithm is relatively stable, with maximum relative energy savings of 12.1 and 6.3% under typical daily operating conditions, and 9.1 and 6.0% during the entire cooling season. In addition, compared to the traditional ES algorithm, the average relative energy savings are 6.4%.
(5) Although the ES algorithm has lower computational efficiency (average calculation time per operating point is close to 10 min), it can still achieve good optimization results compared to traditional heuristic search algorithms (GA and DE algorithms). Therefore, when the system size is small, the optimization target dimension is low, or there are lower requirements for computational efficiency and accuracy, the ES algorithm can still be considered for online optimization of the system.

References

1. Wang J, Zhao T (2022) Event-driven online decoupling control mechanism for the variable flow rate HVAC system based on the medium response properties. Build Environ 218:109104. https://doi.org/10.1016/j.buildenv.2022.109104
2. Wang J, Jia QS, Huang G, Sun Y (2018) Event-driven optimal control of central air-conditioning systems: Event-space establishment. Sci Technol Built Environ 24(8):839–849 https://doi.org/10.1080/23744731.2018.1457410
3. Wang J, Huang G, Sun Y, Liu X (2016) Event-driven optimization of complex HVAC systems. Energy Build 133:79–87. https://doi.org/10.1016/j.enbuild.2016.09.049
4. Wang J, Hou J, Chen J, Fu Q, Huang G (2021). Data mining approach for improving the optimal control of HVAC systems: An event-driven strategy. J Build Eng 39(99):102246. https://doi.org/10.1016/j.jobe.2021.102246
5. Wang J, Zhao T (2024) Medium spatiotemporal characteristics based global optimization method for energy efficiency trade-off issue in variable flow rate HVAC system. Appl Therm Eng, 123132. https://doi.org/10.1016/j.applthermaleng.2024.123132
6. Zhao T, Wang J, Xu M, Li K (2020) An online predictive control method with the temperature based multivariable linear regression model for a typical chiller plant system. Build Simul 13:335–348. https://doi.org/10.1007/s12273-019-0576-7
7. Hou J, Luo X, Huang G, Zhang L, Yu Z, Eftekhari M (2020) Development of event-driven optimal control for central air-conditioning systems. J Build Perform Simul 13(3):378–390. https://doi.org/10.1080/19401493.2020.1737231

Chapter 8
Concluding Remarks

The book discusses the thermal distribution principle and nonlinear optimal control for HVAC water systems in large public buildings. Firstly, the models of the main components in HVAC water systems are introduced, serving as the foundation for subsequent nonlinear optimal control. Then, the most unfavorable loop and thermodynamic stability are examined concerning the thermal distribution principle. To enhance control at the most unfavorable loop and system thermodynamic stability, a differential pressure-based optimal control strategy is presented. Furthermore, the book systematically covers the hysteresis and nonlinear characteristics of the HVAC water system, followed by an exploration of global optimization methods and intelligent optimization algorithms through experimental research. The key findings and contributions of this work are summarized as follows:

(1) Through engineering examples, Chap. 4 in this book points out that the different significances of the most unfavorable loop under operating conditions and design conditions and explains the objective existence of the most unfavorable thermodynamic loop. For two typical forms of air-conditioning water systems, namely continuous regulated air conditioning water systems and on–off regulated air conditioning water systems, the definition and identification methods of the most unfavorable thermodynamic loop are proposed. Regarding the network stability problems caused by the operation and regulation of each branch in variable flow air-conditioning water systems, the concept of thermodynamic stability is proposed, and its evaluation indicators are given.

(2) Based on the definition and identification method of the MUTL, Chap. 5 proposes the optimal control strategy of the DP setpoint based on the maximum valve position loop valve position. The simulation model of the air conditioning water system based on the prototype of the actual project is established, and the optimal control strategy is studied from three aspects of energy-saving effect, condition guarantee, and thermal stability. As the foundation of the DPSPT control strategy, the optimal control for chilled water pumps is also proposed.

T. Zhao et al., *Thermal Distribution Principle and Nonlinear Optimal Control for HVAC Water System*, Indoor Environment and Sustainable Building,
https://doi.org/10.1007/978-981-96-0377-0_8

(3) The HVAC water systems exhibit significant hysteresis and nonlinear characteristics when undergoing optimized control, which is of important reference significance for improving the decoupling optimization effect of the system. Chapter 6 investigates the application methods of spatiotemporal feature models in practical systems from the perspective of practical application. It proposes an online application method based on a dual-loop control architecture. Through experimental studies, the reliability of the spatiotemporal feature model and the online application method has been verified, effectively enhancing the feasibility of the model and method in practical applications. The theoretical guidance and technical support for the online application of decoupling optimization control methods for HVAC water systems are provided.

(4) Traditional fixed-period optimization strategies often result in "untimely" or "unnecessary" optimization actions by ignoring the temporal characteristics of the system, leading to suboptimal decoupling control effects during actual operation. To address this issue, Chap. 7 firstly proposes a variable-period optimization strategy based on a sequential event-driven method (SEDO). Then, comparative experimental studies with traditional time-driven optimization strategies (TDO) validated the SEDO strategy's advantages in decoupling effects, control performance, and energy-saving potential. The SEDO strategy can potentially save approximately 4.0% of daily electricity consumption compared to the TDO strategy, offering better economic benefits. Additionally, the chapter examined efficiency and accuracy issues in solving the global energy efficiency game problem of complex HVAC water systems using traditional optimization methods. By incorporating the system's spatial characteristics, the study thoroughly investigated the global energy efficiency game problem and optimization methods, clarifying the correlation between operating parameters of each device, spatially related operating parameters, and their energy consumption. Besides, based on heuristic search algorithms, an adaptive hybrid differential evolution (AH-DE) algorithm was proposed. Simulation and experimental studies verified the algorithm's reliability in computational efficiency, solution accuracy, and energy-saving potential, effectively addressing spatial coupling issues between devices and control loops.

Compared to traditional TDO strategies, the SEDO strategy demonstrates superior decoupling and control effects, with smaller fluctuations in quality-mass parameters during system operation, making control objectives easier to achieve. The AH-DE algorithm adaptively adjusts the crossover rate and scaling factor based on individual fitness changes, enhancing search accuracy and avoiding local optima. It can save up to 16.1% of electricity consumption under typical daily operating conditions and 17.2% during the entire cooling season.

This book provides essential theoretical guidance and practical reference value for studying global energy efficiency game methods for central HVAC water systems and improving their actual global operational energy efficiency.

The manufacturer's authorised representative in the EU is Springer Nature Customer Service Centre GmbH, Europaplatz 3, 69115 Heidelberg, Germany. If you have any concerns regarding our products, please contact ProductSafety@springernature.com

Printed and bound by CPI Group (UK) Ltd, Croydon, CR0 4YY

30/09/2026

02225265-0011